# <u>Disclaimer</u>

The publisher of this book is by no way associated with the National Institute of Standards and Technology (NIST). The NIST did not publish this book. It was published by 50 page publications under the public domain license.

50 Page Publications.

**Book Title:** Combined Stairwell and Elevator Use During Building Evacuation

**Book Author:** Paul A. Reneke; Bryan L. Hoskins; Richard D. Peacock;

**Book Abstract:** Interest in better understanding of occupant evacuation in buildings is certainly not new. This paper uses the National Institutes of Standards and Technology's (NIST) Egress Estimator software to examine the impact of using both elevators and stairs together in an egress system. The important design parameters for stairs and elevators are examined separately. A method of combining the modeling results for the stairs and elevators to determine an optimal fraction of the population that should be using each system is presented.

**Citation:** NIST TN - 1793

**Keywords:** Egress; Evacuation; Elevators;Stairs

NIST Technical Note 1793

# Combined Stairwell and Elevator Use During Building Evacuation

Paul A. Reneke
Richard D. Peacock
Bryan L. Hoskins

http://dx.doi.org/10.6028/NIST.TN.1793

National Institute of
Standards and Technology
U.S. Department of Commerce

# NIST Technical Note 1793

# Combined Stairwell and Elevator Use During Building Evacuation

Paul A. Reneke
Richard D. Peacock
*Engineering Laboratory*

Bryan L. Hoskins
*Oklahoma State University*

http://dx.doi.org/10.6028/NIST.TN.1793

April 2013

U.S. Department of Commerce
*Rebecca Blank, Acting Secretary*

National Institute of Standards and Technology
*Patrick D. Gallagher, Under Secretary of Commerce for Standards and Technology and Director*

**National Institute of Standards and Technology Technical Note 1793**
**Natl. Inst. Stand. Technol. Tech. Note 1793, 58 pages (April 2013)**
**http://dx.doi.org/10.6028/NIST.TN.1793**
**CODEN: NTNOEF**

This publication was sponsored by the
U.S. General Services Administration, Public Buildings Service

 U.S. General Services Administration

# Table of Contents

# Figures

## Tables

# Combined Stairwell and Elevator Use During Building Evacuation

Paul A. Reneke, Bryan L. Hoskins, and Richard D. Peacock
Fire Research Division
Engineering Laboratory

## Abstract

Interest in better understanding of occupant evacuation in buildings is certainly not new. This paper uses the National Institutes of Standards and Technology's (NIST) Egress Estimator software to examine the impact of using both elevators and stairs together in an egress system. The important design parameters for stairs and elevators are examined separately. A method of combining the modeling results for the stairs and elevators to determine an optimal fraction of the population that should be using each system is presented.

# 1   Introduction and Background

Historically, building egress systems have evolved in response to specific large loss incidents.  This has led to annual expenditures in the billions to install fire protection features, including egress systems, which have little technical basis.  Currently, systems are designed around a concept of providing stair capacity for the largest occupant load floor in the building with little or no consideration of occupant behavior, needs of emergency responders, or evolving technologies.  Aggressive building designs, changing occupant demographics, and consumer demand for more efficient systems have forced egress designs beyond the traditional stairwell-based approaches, with little technical foundation for performance and economic trade-offs. Issues include the design for full-building and phased evacuation of occupants, provisions for the evacuation of individuals with disabilities, and counterflow issues between first responders accessing a building and occupants egressing from a building.

Science-based egress designs have significant potential to mitigate the growing costs of fire protection features in building construction, which are now approximately $63 billion annually in the United States [1]. Adoption of efficient egress designs and advanced egress technologies can therefore have an impact on the life loss due to fire.  For example, improvements in signage, markings, and lighting led to great reductions in egress time from One World Trade Center on September 11, 2001 compared to the 1993 bombing of the same building.  Also, the use of elevators in World Trade Center Building Two (before it was struck by an airplane) led to thousands of lives being saved.

## 1.1   Background
Interest in a better understanding of occupant evacuation in buildings is certainly not new. In 1935, the National Bureau of Standards (NBS) [2] observed stairwells under normal, drill and

laboratory conditions; most were normal conditions. The data were used to suggest possible approaches to calculating minimum width of exits necessary to provide for occupant safety, with the now-ubiquitous 1.3 m (44 in) width stairs deemed to meet the needs of the vast majority of buildings. The London Transport Board [3] observed passengers in nine London subway stations in 1958. They considered a number of parameters that might impact the movement of passengers. Perhaps most widely used and cited, Fruin's 1987 work divides different stair density conditions into six levels of service [4].

More recently, Proulx [5], Lord [6], Peacock, [7] and Hoskins [8] reviewed (largely fire-drill) data on occupant speed, flow, and density. For occupants with mobility impairments, the literature ranges from 0.16 m/s to 0.76 m/s; for studies with no reported impairments, 0.17 m/s to 1.9 m/s. Hoskins [8] reviewed numerous correlations for movement speed in stairwells as a function of occupant density. While there is considerable spread in the correlations and in the underlying data, the correlations center on those included in the Society for Fire Protection Engineering (SFPE) Handbook [9]. This correlation is the basis for the model used for this study.

Like stairwell provisions, the use of elevators during fire emergencies has been considered as well. As early as 1914, properly protected elevators were seen as essential in taller buildings, but automatic elevators were deemed unsuitable in the 1935 NBS report [2]. More recently, there has been considerable attention to the use of elevators to speed up building evacuation. This included studies of the feasibility of elevator evacuation, human behavior, and the use of elevator lobbies as areas of refuge [10]. Protection from heat, flame, smoke, water, overheating of machinery, and loss of electrical power were seen as important to elevator design [10]. Using model calculations for example buildings, elevator evacuation was estimated to speed evacuation by 16 % to 25 % compared to evacuation by stairs alone; the taller the building the greater the impact.

Based on ongoing research sponsored by the U. S. General Services Administration, Public Buildings Service (GSA/PBS), building codes are beginning to recognize the use of elevators for occupant self-evacuation after decades of training occupants that elevators are not to be used during fire evacuations. With this change in evacuation procedure, it is important to provide first responders and all building occupants, including those unable to negotiate the stairs, with an understanding on the safe use of elevators during emergencies.

## 1.2   Current Study

As part of this larger project to understand stair and elevator usage for occupant evacuation during fire emergencies, this report provides an analysis of potential trade off decisions that will face building designers and operators with the combined use of elevators and stairwell during fire emergencies through the use of egress modeling techniques to evaluate the relative impact of a range of design and operational parameters on building evacuation.  Parameters studied include

- Building capacity and building height,
- Relative elevator and stairwell capacity during emergency egress,

- Full building evacuation versus partial evacuation of effected floors
- The potential for reducing exit stair capacity when installing occupant evacuation elevators as well as the potential for reducing the number of exit stairs depending on various parameters (e.g., building height, number of occupants, occupant demographics, etc.)

While the software used calculates a large number of variables that show where evacuees are in the building at any given moment during the simulation, to keep the analysis tractable this paper will focus on the time to total evacuation of the building or egress time. In a full design, especially given a fire that creates hazardous conditions, times to vacate a particular floor or section of the building can be important in addition to total egress time for the full building.

This report is broken into a number of sections. Section 2 describes the basic features of the Egress Estimator [11] and the parameters studied. The next two sections discuss the individual sub-models. First the sub-model of stairwell evacuation is examined, focusing on how input parameters change the results. Next, the elevator sub-model is presented. In section 5, a method of combining the results of these two sub-models is presented. The discussion, summary and possible avenues for new research follow.

# 2   Study and Modeling Procedures

In this section the basic features of the model, the Egress Estimator [11], and the parameters studied will be discussed. The next two subsections describe first the stairwell model and then the elevator model.

## 2.1   Stairwell Modeling

The stair model is built on the theory developed in the SFPE handbook [9], using the same simplified building model and the speed versus density correlation to build a 1D partial differential equation (PDE) model of stair evacuation.

The Egress Estimator assumes a specific building and stair geometry. Occupants are evenly divided on all the floors. For each stair on each floor there is a corridor. The evacuees travel the corridor and enter the stairs on a landing. Between the floors is a user defined number of flights of stairs and landings, the riser height and number stairs then defines the height between the floors. Evacuees entering from a floor are assumed to merge evenly with those in the stairs. At the bottom of the stairs on the first floor is an exit hallway leading to the exit door. While the Egress Estimator can model occupants on the first floor entering the stairs, merging with those coming down the stairs and exiting the building in all of the simulations examined in this study the first floor is assumed to evacuate through different exits.

The Egress Estimator tracks where people are and how fast they are traveling by dividing the travel path into a number of cells. The two state variables of the PDEs used are density and speed.

The density equation is

$$\dot{D} = \frac{1}{W}\frac{\partial f}{\partial x} \tag{1}$$

where $\dot{D}$ is the rate of change of density, $W$ is the effective width of the cell, $f$ is the flow of evacuees and $x$ is the length in direction of the flow. The effective width is the measured width minus boundary zones that describe how close people get to the walls. For the Egress Estimator, boundary zones are assumed to be 150 mm on each side of the stair for a total of 300 mm.

The speed equation is

$$\dot{s}_j = \begin{cases} a\left(s_{max} - s_j\right) & \text{for } D \le D_{min} \\ a\left(\dfrac{D_{max} - D_{j+1}}{D_{max} - D_{min}}\right)s_{max} - s_j & \text{for } D > D_{min} \end{cases} \tag{2}$$

where $j$ is the index of the cell with the flow of people traveling from the $j^{th}$ cell to the $j+1$ cell, $s$ is the speed, $s_{max}$ is the maximum speed in the cell, $D_{min}$ is the density below which the density

4

does not affect speed, $D_{max}$ is the maximum density a cell can have, $a$ is a parameter that controls the speed a cell equilibrates. The speed equation uses the density in the $j+1$ cell with the inherent assumption that conditions in front of an evacuee controls their speed. If the density is low enough, speed is controlled by a predefined maximum speed occupants can travel.

The stairwell model was compared to four fire drill evacuations from an ongoing project at NIST to collect stair movement data during fire drill evacuations [12]. A summary of the results is given in Figure 1. In the graph, the evacuation time for each evacuee for the four stairwells is normalized by the total building egress.

**Figure 1. Estimated egress time compared to measured times in several fire drill evacuations. Individual times, both calculated and measured, are normalized by the total egress time.**

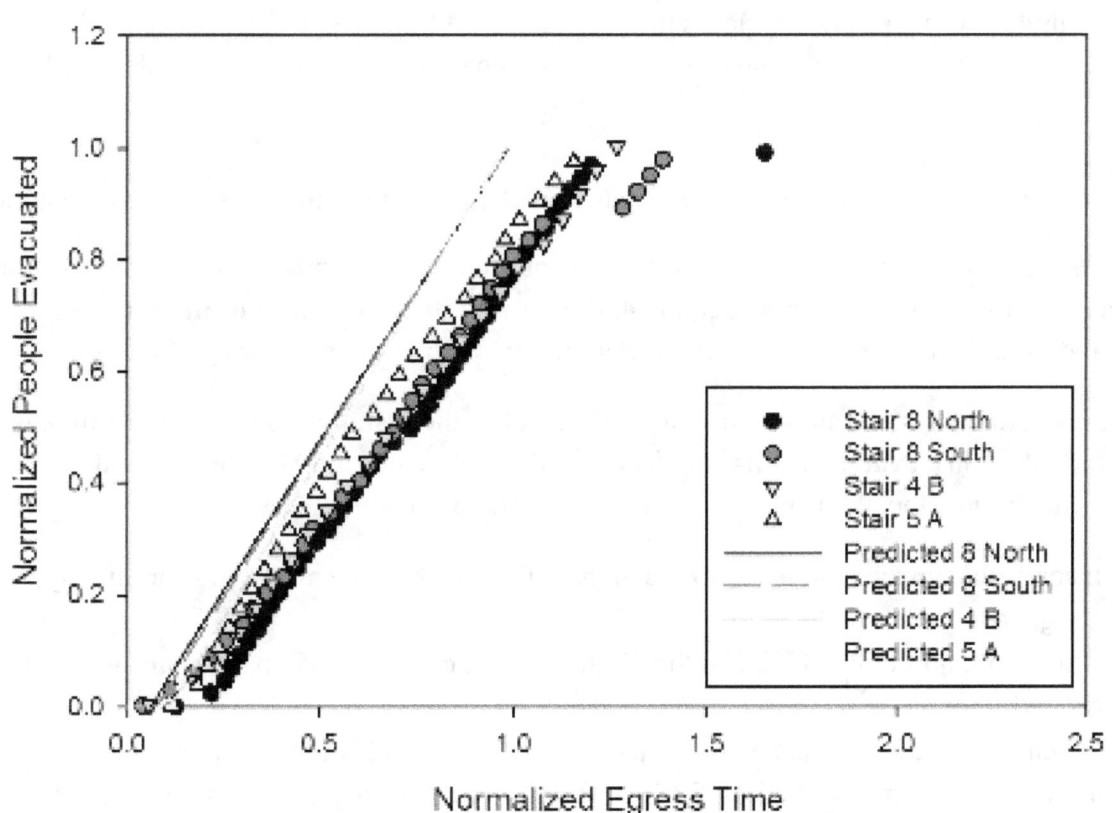

Predicted egress times shown in Figure 1 tend to somewhat underpredict those from experimental data. In the four evacuations studied, at least 98 % of all occupants in each of the evacuations were out of the building by the predicted evacuation time plus 30 %. In only one evacuation did a single person (or less than 0.3 % of the total population) take more than predicted egress time plus 100 %. This is not surprizing since the underlying SFPE correlation is not intended to estimate evacuation time for individual stragglers. Estimating the egress time of

people that significantly delay evacuating requires a human behaivor model. For simple
analyses, a design margin can be added to account for these stragglers.

## 2.2 Elevator Modeling

The elevator modeling is based on Klote's model in ELVAC [13]. The main equation in ELVAC
is

$$t_e = \frac{1+\eta}{J} \sum_{r=1}^{R} \sum_{n=1}^{N_r} t_{r,n} \tag{3}$$

where $t_e$ is the total egress time, $\eta$ is the overall inefficiency that increases the egress time above
the ideal, $J$ is the number of elevator cars, $R$ is the number of floors, $N_r$ is the number of trips that
have to be made to the $r^{th}$ floor to clear it and $t_{r,n}$ is the time the $n^{th}$ trip to the $r^{th}$ floor takes.

The Egress Estimator calculates the $t_{r,n}$ exactly the same as ELVAC. The model takes a user
defined maximum velocity and acceleration along with the distance to the $r^{th}$ floor calculated
from the number of steps between floors to calculate travel time to and from the floor. It
calculates the door open and closing times and load/unload times are exactly as calculated by
Klote.

However, instead of summing the trips and using an efficiency parameter, $\frac{1+\eta}{J}$, to adjust the
total egress time, the Egress Estimator uses an elevator control algorithm that accounts for this
inefficiency. Still, as will be seen in section 4, the form of this equation is useful for
understanding the impact of various design changes on the overall elevator egress time.

The Egress Estimator's elevator control algorithm makes the assumption that if there are any
people left when an elevator car fills and departs a floor, they instantly hit the call button for the
next car. In addition, the algorithm has the following three rules[1]:

- It sends the next available car to the highest floor that has a call and no car already
  assigned
- If there are no floors with a call that do not have a car assigned, the elevator waits at the
  ground floor
- When an elevator has picked up all passengers on a floor, if the elevator load is less than
  a set amount (with a default of 80 % of the elevator capacity), it will stop to pick up

---

[1] Note that this algorithm is similar but not identical to the current requirements of ASME A17.1
for elevator evacuation. Cars are sent to the highest effected floor, cars wait at the lowest effect
floor for additional calls (here based on full-building evacuation), and the default loading
capacity of 80 % to 100 % is consistent with the ASME requirements. ASME requirements do
not, however, include provisions to stop at additional floors. The impact of elevator loading
capacity, placement, and movement is discussed in section 4.

additional passengers. The car stops at the next floor down that has a call and no car
assigned. If a car is assigned to every floor with a call, the car proceeds to the ground
floor.

The control algorithm is used for three reasons. Firstly, it eliminates the need to use an efficiency parameter $1 + \eta$. Secondly, it allows the elevator evacuation to evolve in time in concert with the stairwell evacuation. Finally, it provides for future expansion to evaluate different strategies of both design and control. The algorithm has been compared to an example from reference [13]; the results were within 1 % of the answers using ELVAC. All of the parameters could be varied to study the impact of different choices for the basic algorithm.

## 2.3 Study Parameters[2]

To represent a number of different buildings that could use combined stairwell and elevator egress systems, a range of parameters were varied. The parameters were applied to just the stairwells, just the elevators, or both.

### 2.3.1 Stairwell Parameters

Six different stair widths were used in the modeling. They were four typical widths, 1118 mm (44 in), 1219 mm (48 in), 1422 mm (56 in) and 1524 mm (60 in), as well as two non-typical widths, 1270 mm (50 in) and 1905 mm (75 in). The two non-typical widths arise from two limiting requirements in the 2012 edition of the Life Safety Code [14]. The first requires that there is 7.6 mm of stair width per person per floor[3]. There is also a code requirement of at least three means of egress for occupant loads more than 500 people per floor, but not more than 1000 people per floor[4] [14]. For a population of 500 people, the total required width is 3800 mm. For a building at the transition point between being required to have two or three stairwell shafts, both three 1270 mm (50 in) wide stairs and two 1905 mm (75 in) wide stairs would be code compliant.

The four typical stair width cases were run with both 2 and 3 stairs for comparisons. The 1905 mm (75 in) width was run only with two stairs while the 1270 mm (50 in) wide was only run with three stairs.

In all simulations, the riser height was 178 mm (7 in) and the tread depth was 279 mm (11 in). Furthermore, there were two flights of stairs per floor with each flight having 9 steps so the height of a floor is 3.2 m (10.5 ft) high. These values, taken as typical of building construction, were consistent with those cited by the authors mentioned in section 1.1 [5-8].

---

[2] The selection of values for any particular parameter is not intended to be indicative of how proper design should be conducted. Instead, it is intended to compare a range of values. In practice, values for a given parameter could fall outside of this range.
[3] From Table 7.3.3.1 of the Life Safety Code for "All others." [14] The 0.3 in requirement was used which results in a slightly greater width.
[4] From section 7.4.1.2(1) of the Life Safety Code [14].

### 2.3.2 Elevator Parameters

A range of elevator accelerations from 1.0 m/s$^2$ to 3.8 m/s$^2$ were included with a default acceleration of 1.0 m/s$^2$. Likewise, a range of maximum speeds for the elevators from 3.0 m/s to 11 m/s were condsidered; the default maximum speed for elevator cars was 3.0 m/s. For zoned elevators, a range of maximum speeds is used for the upper zones, with a default speed of 8.0 m/s. Center opening, 1219 mm (48 in) elevator doors were assumed.

### 2.3.3 Building Parameters

The selected building heights for stair and elevator evacuations are 22 m (7 stories, 74 ft), 112 m (35 stories, 368 ft) and 160 m (50 stories, 526 ft). These were chosen simply to provide a range of building heights to best demonstrate the functional relationships that are being explored. The elevator evacuations used buildings of other heights, 48 m (15 stories, 158 ft), 67 m (21 stories, 221 ft) and 80 m (25 stories, 263 ft), to examine certain phenomena more closely. When testing the effect of elevators in a phased evacuation, a section of five floors are evacuated to the floor below.

### 2.3.4 Building Population

Unless stated otherwise, a population of 504 persons/floor was used because this is the point (actually slightly above the point), at which the Life Safety Code requires a third stair shaft to be added. The reason for using 504 instead of 500 persons/floor is that 504 can be divided evenly between the stairs and elevators for all of the scenarios used. As will be seen in the analysis, egress times for lower or higher floor populations are linearly related so this chosen value for the initial analysis does not limit the applicability of the results.

The percentage of occupants using the stairs varied in the different scenarios. The percentage of the occupants using the elevators was 0 %, 25 %, or 50 %. The remainder of the population was evenly distributed to the different stair shafts.

In the phased evacuation cases, the incident floor, two floors above and two floors below are evacuated. To simulate this in the Egress Estimator, a 6 story building was used. The incident floor is the third floor although no priority is given to the population on that floor. Noting that once occupants are below the zone in either the stairs or the elevators, the difference in egress time is proportional to the total additional distance traveled so that regardless of building height, the egress time differs only by a constant value for a single given building.

# 3  Stairwell Evacuation

For the full-building evacuation using stairs, scenarios were examined by varying the height of the building, the percentage of the population using the stairs, the number of stairs, and the stair widths.

## 3.1  Full-Building Stairwell Evacuation

In the SFPE Handbook [15], Pauls provides three simple correlations for the evacuation time of a building based on the number of people using a stair and the effective width of the stair. The equations have two forms

$$T = A + Bp^{0.73}$$
$$T = A + Bp$$
(4)

where A and B are determined from evacuation data for types of buildings and occupancies and $p$ is given by

$$p = \frac{\text{Total number of people}}{\text{Effective width of stairs}} = \frac{P/S \cdot R}{W}$$
(5)

where $P$ is the population per floor, $S$ is the number of stairs, $R$ is the number of floors and $W$ is the effective width of the stairs.

To examine the impact of each variable on the total egress time first note from eq. (4) and (5) that the total egress time is linearly proportional to the number of floors and the population per floor and inversely proportional to the number of stairs and the effective width of the stairs as shown in eq. (7).

$$T = A + B\frac{P/S \cdot R}{W} = A + B\frac{PR}{SW}$$
(6)

If eq. (6) holds it says the rate of flow in the stairs is what determines the flow out the building. For that to be true requires that the flow of evacuees in the other parts of the building are faster than the flow in the stairs.

The rest of this section will look at how well the Egress Estimators stair model is represented by the linear form in eq. (4). First the relationship of the population to the egress time will be examined and then the effective stair width on egress time. Finally the interaction of population and effective stair width will be examined.

### 3.1.1 Effect of Total Population in Stair on Total Egress Time

The relationship of total egress time to each parameter that determines total population in the stair will be examined individually first then combined to evaluate the impact of all the parameters governing the total population on the total egress time.

#### 3.1.1.1 Building Height

As can be seen in Figure 2, the relationship between the time required to evacuate by stairs, as calculated by the model, is nearly linear with respect to the building height. This pattern holds for the instances when different fractions of occupants use the elevators.

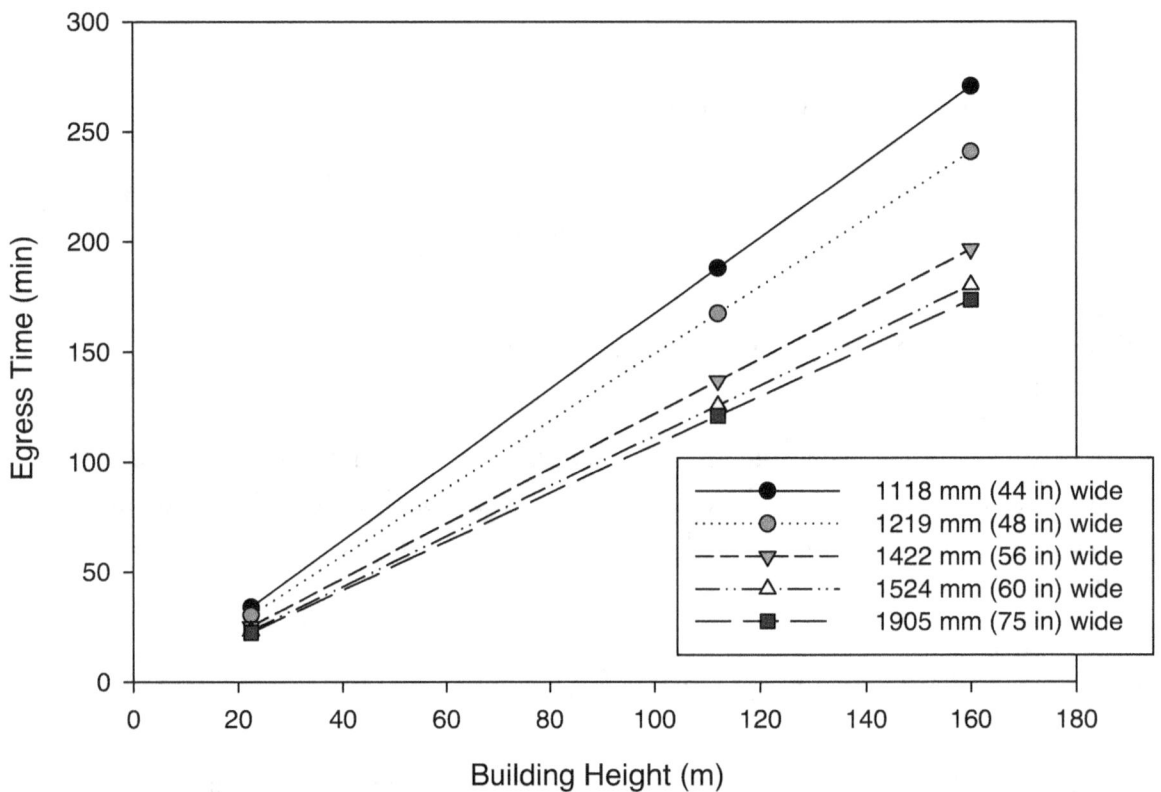

**Figure 2. Estimated stair egress time for a range of building hights and stair widths**

#### 3.1.1.2 Percentage of Population Using Elevators

As was the case with the building height, the model calculates a nearly linear relationship between the percentage of the population in the stairs and the required egress time by the persons in the stairs. Figure 3 shows the relationship between egress time within the stairs and the percentage of the population using the stairs in the 160 m (50 stories, 525 ft) building with different stair widths. The other buildings show similar trends.

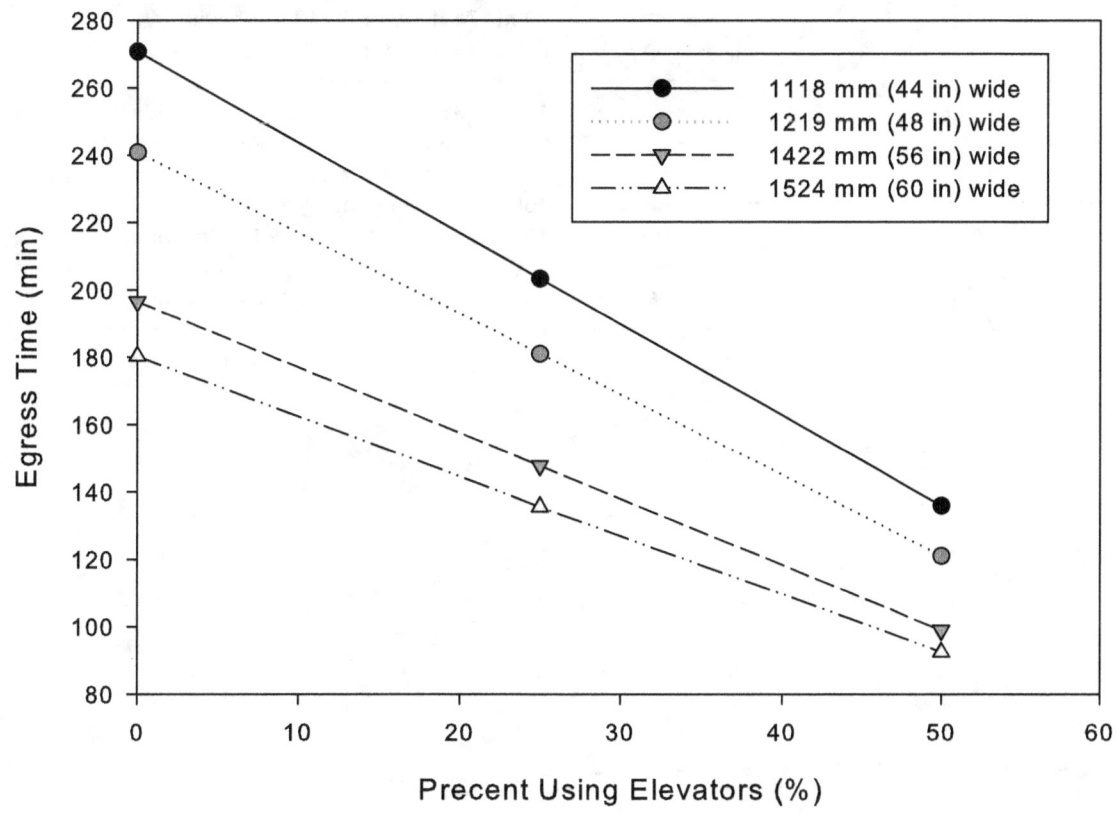

**Figure 3. Estimated egress time as a function of building population using the stairs and of stair configuration.**

### 3.1.1.3 Number of Stairs

For the 1270 mm and 1119 m wide stairs, the number of stairs was varied from two to three. As shown in Figure 4, for the stair only condition, the scenarios with two stairs required more time for safe egress than the three stair scenarios. The same trend was seen in the scenarios where a percentage of the population was assigned to the elevators. The trend in each building is the same. Listing from the slowest to the fastest: 2 stair 1119 mm wide stairs, 2 stair 1270 mm wide stairs, 3 shaft 1119 mm wide stairs and finally the fastest is 3 shafts 1270 mm wide stairs. As would be expected, this is the same order as the total cumulative stair width in each building.

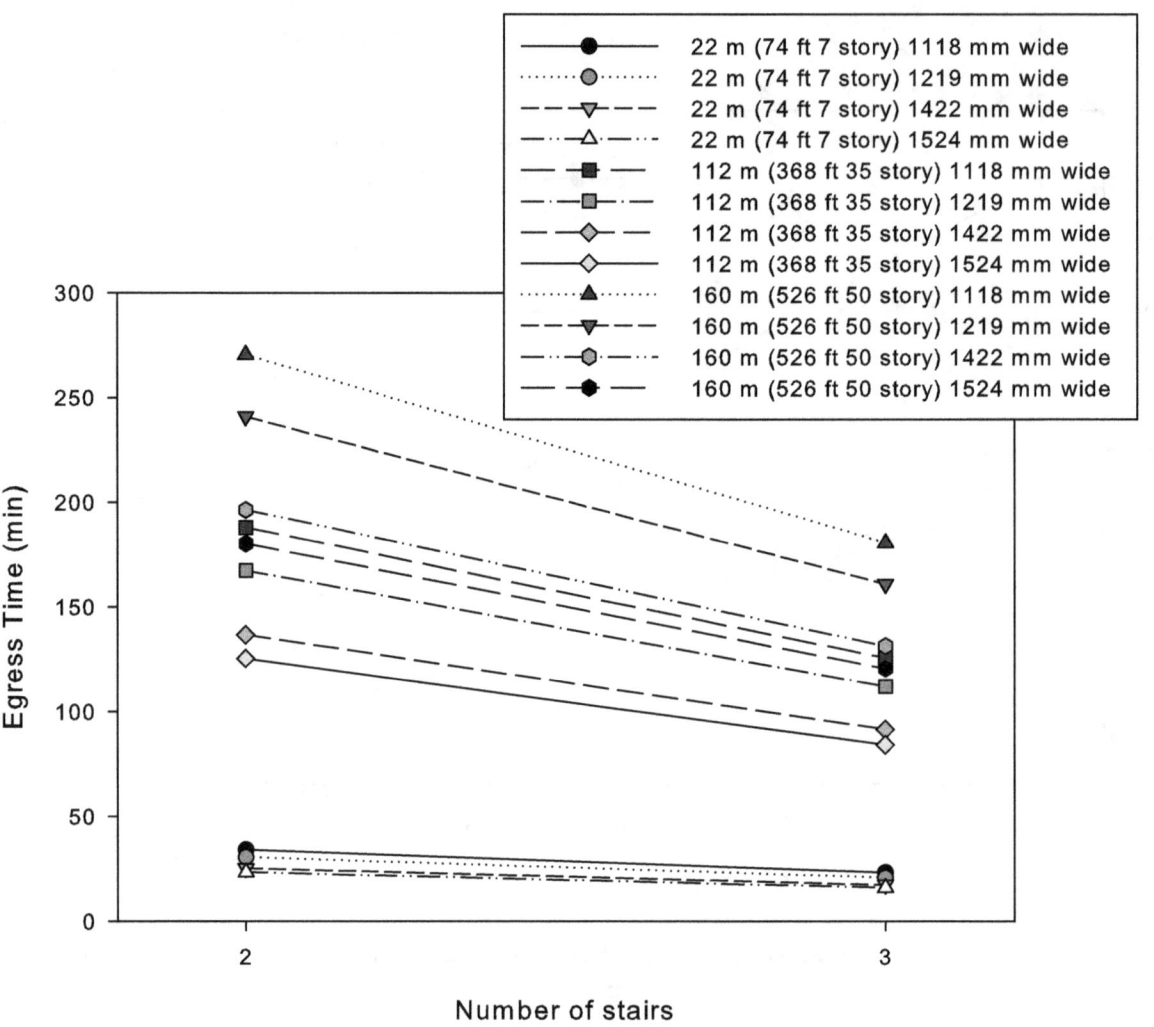

**Figure 4. Relationship between number of stairs and estimated egress time in stairs. All buildings have 504 people per floor.**

Note that evacuation of the 112 m tall building with stair widths of 1117 mm (44 in) is slower than the 160 m tall building with 1524 mm stairs. The explanation, using eq. (7), will be discussed later in section 3.1.3.

### 3.1.1.4 *Total population effect*
Individually, the parameters seem to have the correct relationship. In Figure 5 the total egress time is grouped by the width of the stair. As can be seen the relationship between the number of people using the stair and the egress time is linear for each stair width, consistent with the form in eq. (7).

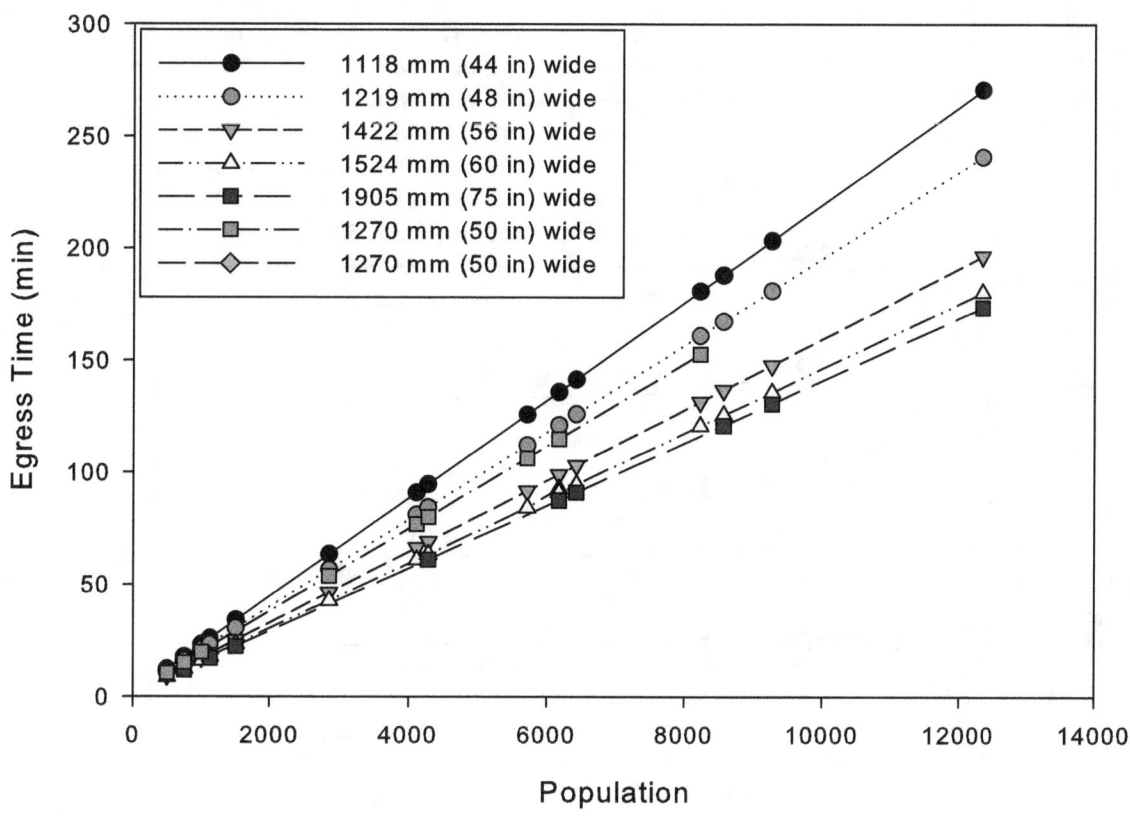

**Figure 5. Total egress time for different stair widths against total population using the stair**

### 3.1.2 Effect of Stair Width on Total Egress Time

From eq. (6), the total egress time should be inversely proportional to the effective width of the stairs. Figure 6 shows the impact of effective stair width on the total evacuation time when the total population using the stair is held constant.

The first 4 points for each group has the classic shape of $y=1/x$ but the last point doesn't fit. As stated above, the model underlying the correlation is that the controlling point in the evacuation is the stairs. The effect of the stairs on egress time is controlled by the width of the stairs. However, any building has walls, doors, and other potential controlling factors. If the stairs become wide enough, other factors in the buildings can become more important. If the stairs are sufficiently wide such that flow travels at the maximum speed then adding additional width would not speed up the flow. For the base design used in this study, 1905 mm wide (75 in) stairs appear to be wide enough that another feature in the building design is serving as the primary restriction on flow.

13

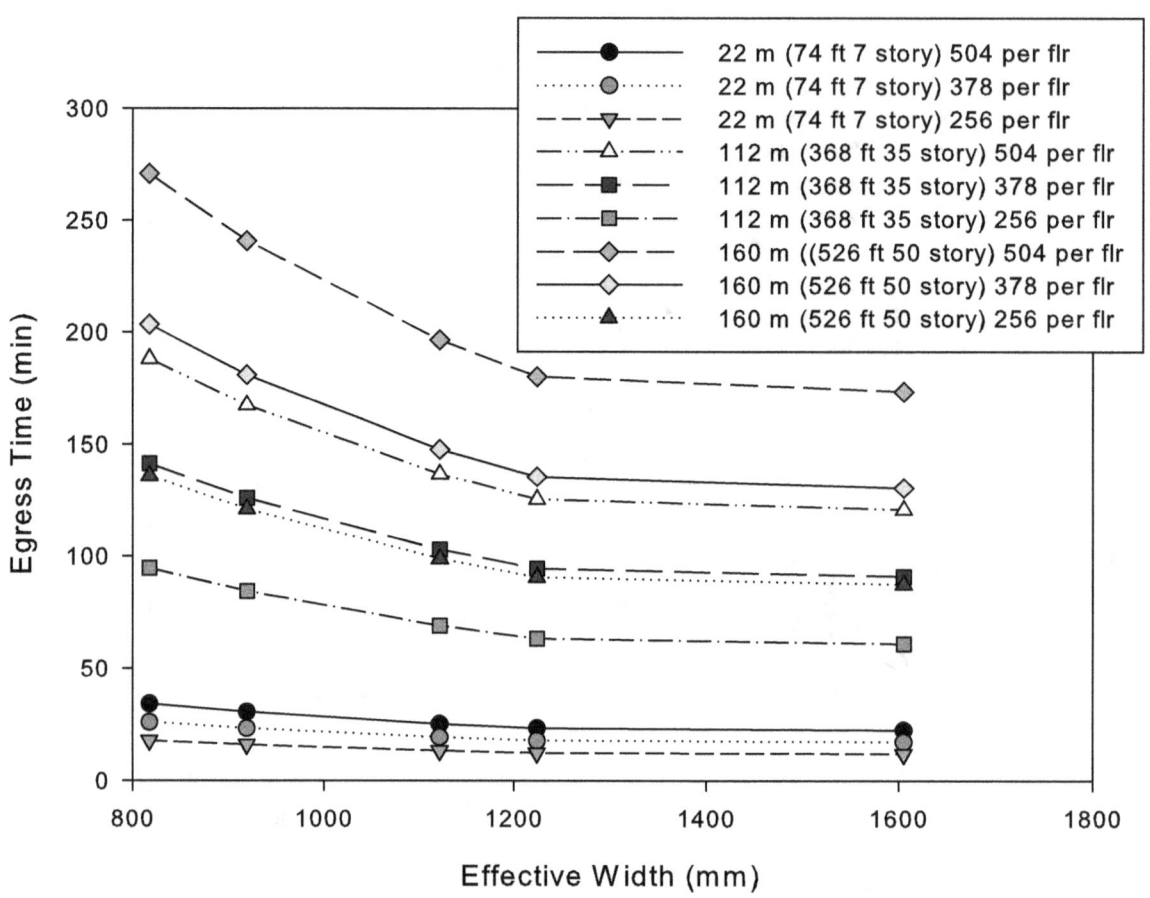

**Figure 6 Effect of width of stair on total evacuation time**

### 3.1.3 Combined Impact on Total Egress Time of *p*

The components of the parameter *p* (total building population / effective stair width) in the linear model of total evacuation time have the expected relationships. So the last question is does the Egress Estimator predict a linear relationship for a set of buildings that are identical except for the total population using each stair and the width of that stair.

In Figure 7 the data gives two lines. For effective stair widths less than 1225 mm (48 in), there is one relationship between *p* and the total egress time. For stairs with an effective width of 1605 mm (63 in), which is the 1905 mm (75 in) case, other aspects of the basic building design limit the impact of widening the stairs, but still with a linear relationship.

14

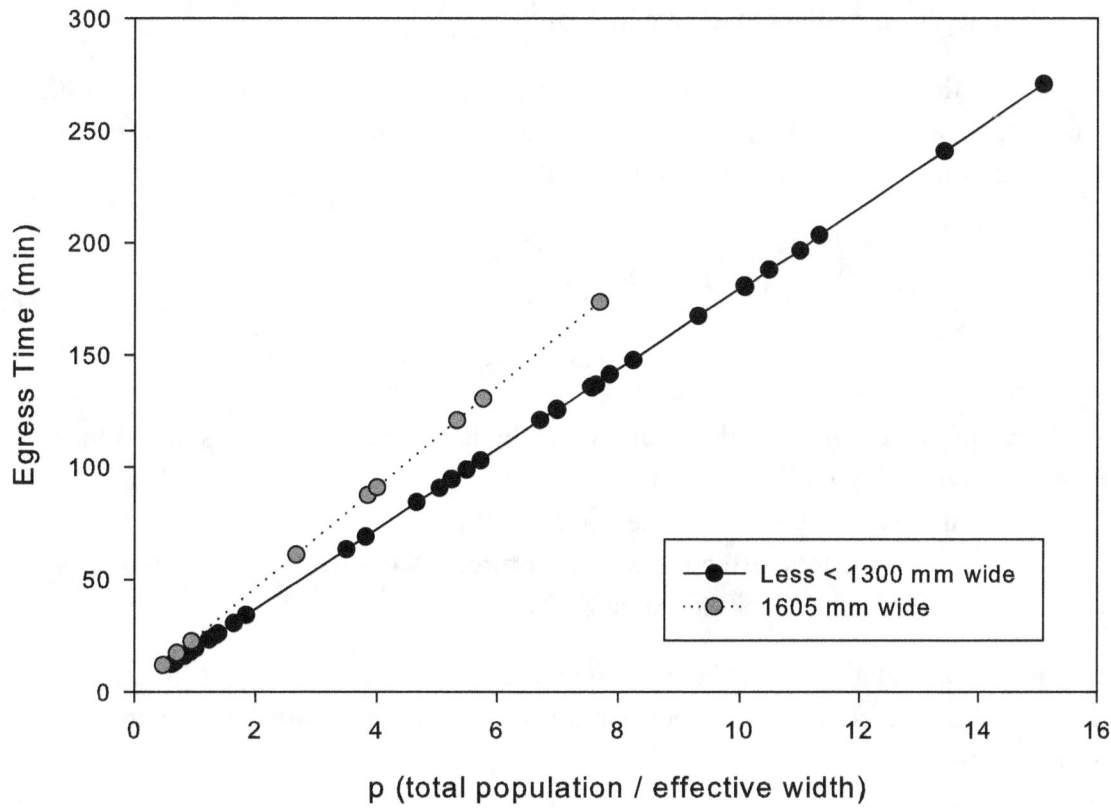

**Figure 7. Total egress time with respect to parameter p.**

It was noted in Figure 4 that 112 m tall building with stair widths of 1117 mm (44 in) is slower than the 160 m tall building with 1524 mm stairs with a significantly larger population. From eq. (6), values for the effective width of a stair in the taller building will allow the total egress time will be smaller when

$$p_s = \frac{P_s}{w_s} > \frac{P_l}{w_l} = p_l$$

$$w_l > \frac{P_l}{P_s} w_s$$

(7)

where the subscript $s$ stands for the smaller building or building with the smaller population and $l$ stands for the taller building, $P$ is the total population in the building and $w$ is the effective width of the stairs. With values for the two buildings in question and the 160 m tall building will have a shorter egress time as long as the stair width is greater than 1478 mm.

15

This relationship can be seen graphically in Figure 5. If a horizontal line is drawn for a particular maximum egress time the wider stairs will have total egress times below the maximum egress time for larger populations than stairs that have less width.

The case of providing the same total stair width using 2 or 3 stairs is examined using two 1905 mm (75 in) stairs against three 1270 mm (50 in) wide stairs. If eq. (6) used total width instead of effective width, then $p$ for the two cases would be the same when

$$\left( \frac{\frac{P}{2}}{\frac{w_T}{2}} \right)_{\text{Two stairs}} = \frac{P}{w_T} = \left( \frac{\frac{P}{3}}{\frac{w_T}{3}} \right)_{\text{Three stairs}} \tag{8}$$

where $P$ is the total population in the building and $w_T$ is the total width of all the stairs. When considering the effective width, which is equal to the *w-300 mm* where *w* is the width and the 300 mm is the total of the boundary layers in the stairs. So the total effective width for two 1905 mm stairs is 3510 mm while the total effective width for three 1270 mm stairs is 2910 mm. So eq. (7) says that the two 1905 mm stairs should be faster.

The Egress Estimators calculations give the opposite results to eq. (6). The model finds that stairs 1905 mm wide reach a point where other features of the building start to have significant impact on the egress time.

## 3.2  Phased Stairwell Evacuation

The phased evacuation consisted of five floors, with 504 persons per floor that were to be evacuated. All occupants from those five floors exited to the floor immediately below these floors; the equivalent of evacuating the upper five floors of a six floor building.

These results are shown graphically in Figure 8.

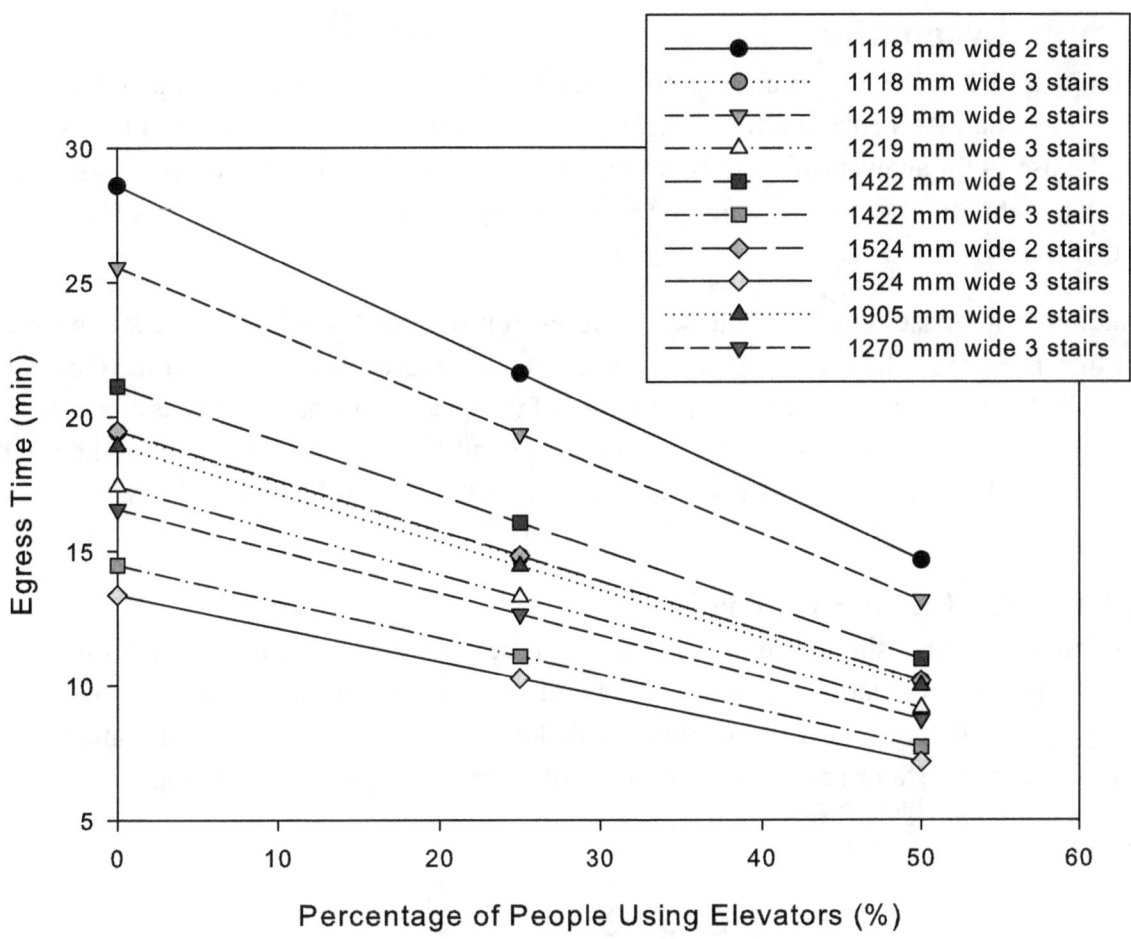

**Figure 8. Phased evacuation relationship between stairs and fraction of population using stairs**

As had been the case for the complete building evacuation, there tends to be a linear relationship between the total population using the stairs and the amount of time required for occupants to complete their evacuation.

# 4 Elevator Evacuation

This section will look at the impact on elevator egress of different parameters. As mentioned in section 2.2, Klote defines a equation for a single bank of elevators evacuating a set of floors, Eq. 3, that can be useful for analysis of a full building evacuation. While the equation is not used in the Egress Estimator, the form of the equation is used and provides for simplifications for the analyses that follow.

The elevator algorithm uses the following simple rules. It is assumed that if there are still people on a floor they press the call button as soon as an elevator car closes its doors. After the elevator cars are recalled to the ground floor at the beginning of the evacuation, each car is assigned to the highest floor with an active call button that has not already had a car assigned. If all the floors with active call buttons have a car assigned the car waits at the ground floor until a new call is activated.

## 4.1 Full-Building Elevator Evacuation

First, a full building with a single bank of elevators will be examined. To make the analysis easier, assume that the population on the floor is an integer multiple of the maximum capacity of the cars, say $N_B$, and that every car is completely filled on every trip. Furthermore, assume that since every trip to the $r^{th}$ floor has the same number of passengers and that every trip takes the same amount of time, $t_r$. This simplifies a part of eq. (3) as follows

$$T_B = \sum_{r=1}^{R}\sum_{n=1}^{N_r} t_{r,n} = \sum_{r=1}^{R} N_B t_r = N_B \sum_{r=1}^{R} t_r \qquad (9)$$

where $T_B$ can be thought of the amount of time it would take to evacuate the building using only one car.

### 4.1.1 Population

Consider the number of trips it takes to evacuate a floor, $N_B t_r$. If the population on a floor is reduced by one car load, the time it takes to empty that floor reduces by $t_{r,k}$ to get $(N_B - 1)t_r$. Reducing the population by 2 car loads reduces the time it takes by $2t_{r,k}$ to get $(N_B - 2)t_r$. For large enough changes in population per floor, on the order of an elevator carload, the evacuation time will reduce linearly.

Figure 9 shows the egress times of using 8, 10 or 12 580 kg (1280 lb) cars in three building sizes: 48 m (157 ft, 15 stories), 80 m (262 ft 25 stories) and 160 m (525 ft 50 stories). The independent variable is the fraction of the 504 people per floor that use the elevators for evacuation. All 9 cases are linear with respect to the population per floor using the elevator cars.

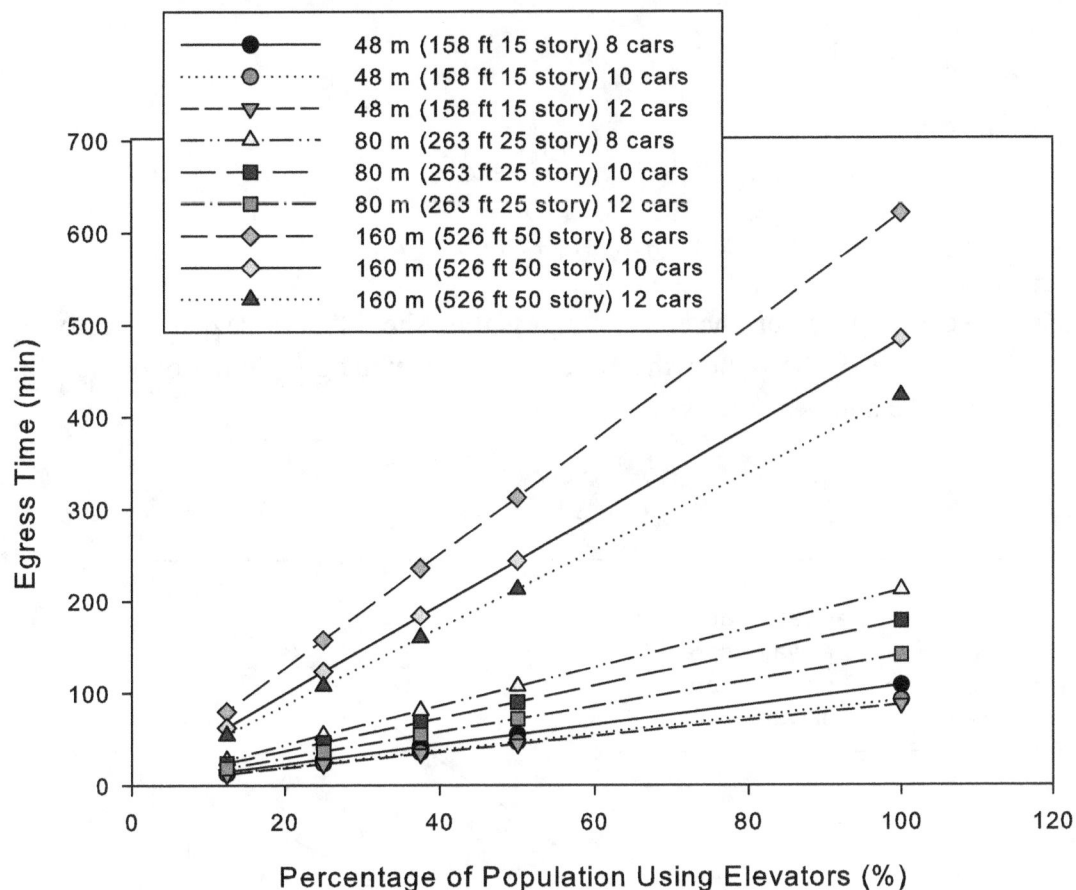

**Figure 9. Impact of a change in population using the elevators on estimated egress time for a range of building heights and elevator configurations.**

### 4.1.2 Number of Floors

Remember that $T_B = \sum_{r=1}^{R_B} N_B t_r$ is the time it takes to evacuate a building with a single car with perfect efficiency that has the number of floors, $R_B$. Now assume that $R_B$ is such that elevator cars only reach maximum speed at the $R_B$ floor. If $t_B$ is the time it takes to make a round trip to the $R_B$ floor and load and unload a full load of passengers and $\Delta t$ be the amount of time it takes a car to go one floor at maximum speed, then $T_{B+1} = T_B + N_B(t_B + \Delta t)$ is the time it would take to evacuate a building one story taller. For a building $R_B + 2$ stories taller, the evacuation time would be $T_{B+2} = T_B + N_B(t_r + \Delta t + t_r + 2\Delta t)$. For a building with $R_B + \Delta r$ floors, the elevator egress time is

$$T_{B+\Delta r} = T_B + N_B\left(t_B + \Delta t + t_B + 2\Delta t + \cdots + t_B \Delta r \Delta t\right)$$

$$= T_B + N_B\left(\Delta r t_B + \Delta t \sum_{r=1}^{\Delta r} r\right) \qquad (10)$$

19

where $\sum_{r=1}^{\Delta r} r = \dfrac{\Delta r(\Delta r + 1)}{2}$ so that

$$T_{B+\Delta r} = T_B + N_B\left(\Delta r t_B + \frac{\Delta r + 1}{2}\Delta t\right) \tag{11}$$

$$= N_B \Delta t \Delta r^2 + N_B(t_B + \Delta t)\Delta r) + T_B$$

Eq. (11) says that above a certain floor, the elevator egress time should increase quadratically with number of floors. Figure 10 shows how the increase in estimated egress time follows a quadratic with respect to the number of floors.

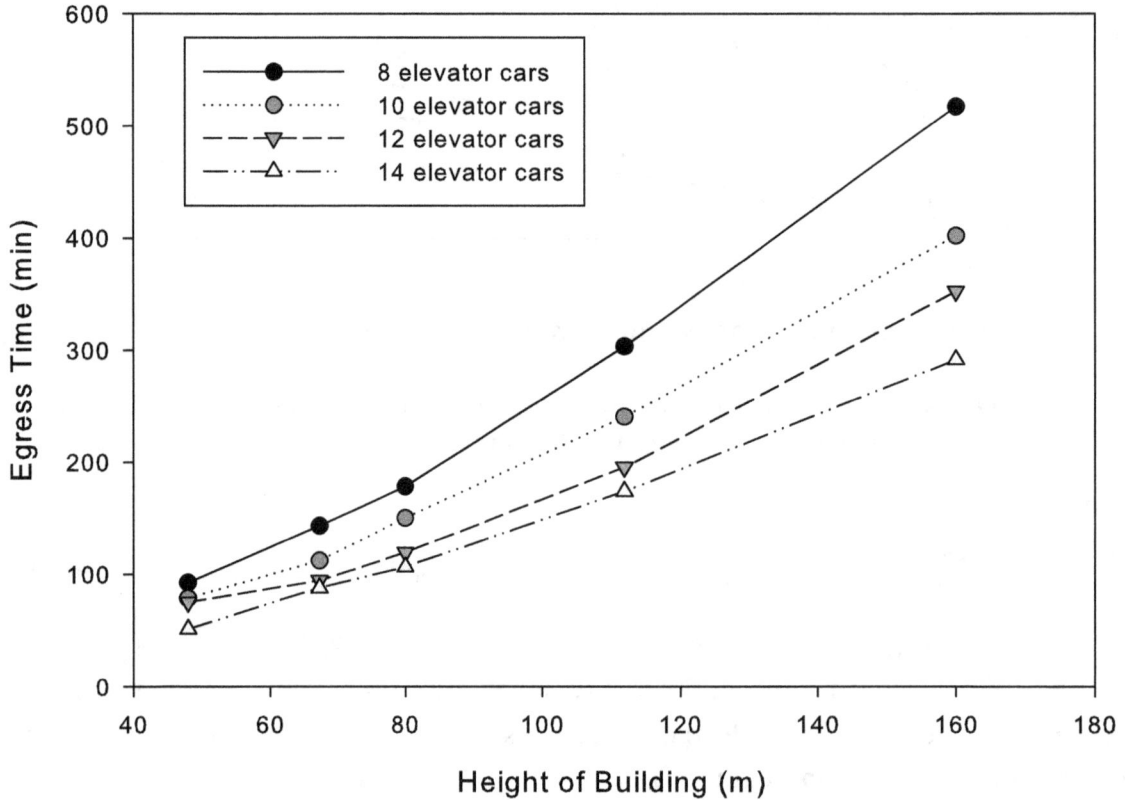

**Figure 10. Impact of increasing the number of floors on estimated egress time for a range of elevator configurations.**

### 4.1.3   Number of Cars

To look at the impact the number of cars has on a building it is first useful to combine eq. (3) and (10) to give

$$t_e = \frac{1+\eta}{J}T_B \tag{12}$$

20

For a given building, $T_B$ is constant. Assuming that $\eta$ is also constant for a given configuration, the total egress time should be inversely proportional to the number of elevator cars, $J$. As can be seen in Figure 11, the Egress Estimator gives the same inverse relationship.

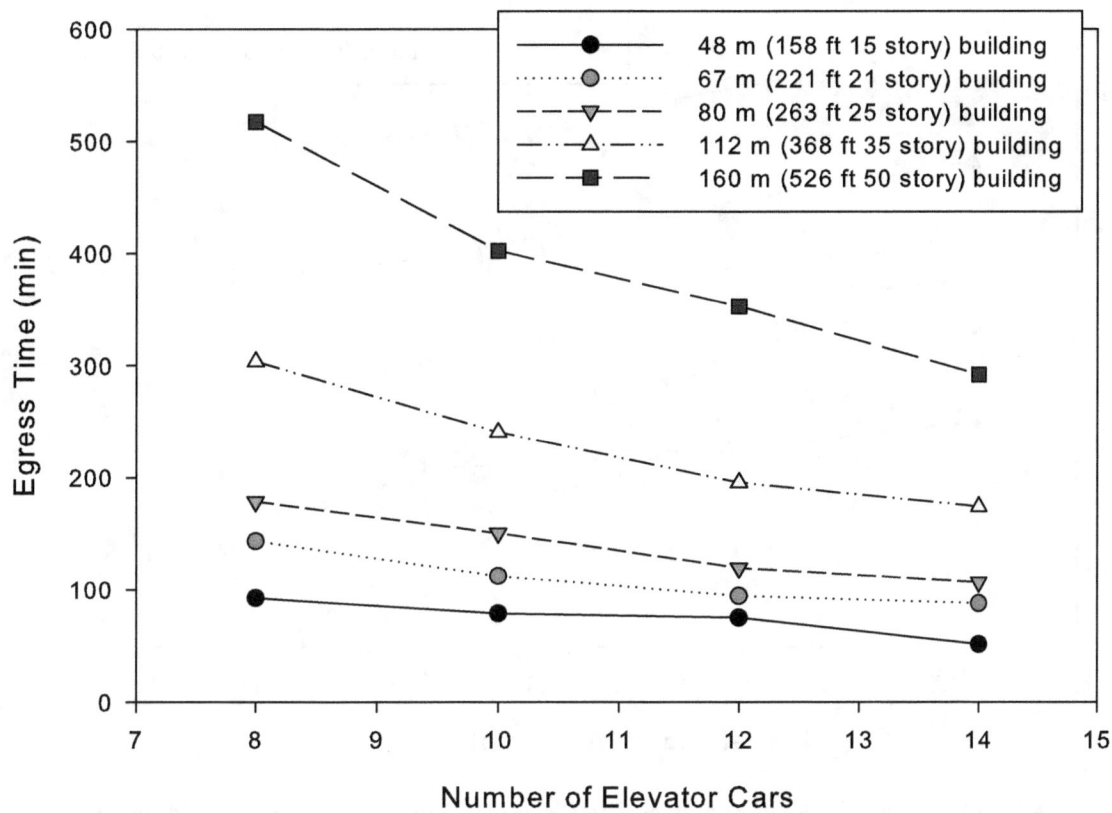

**Figure 11. Impact of additional elevator cars on egress time for a range of building heights and elevator configurations.**

It is easier to see the relationship in a log-log plot[5] where the inverse relationship should be linear as seen in Figure 12.

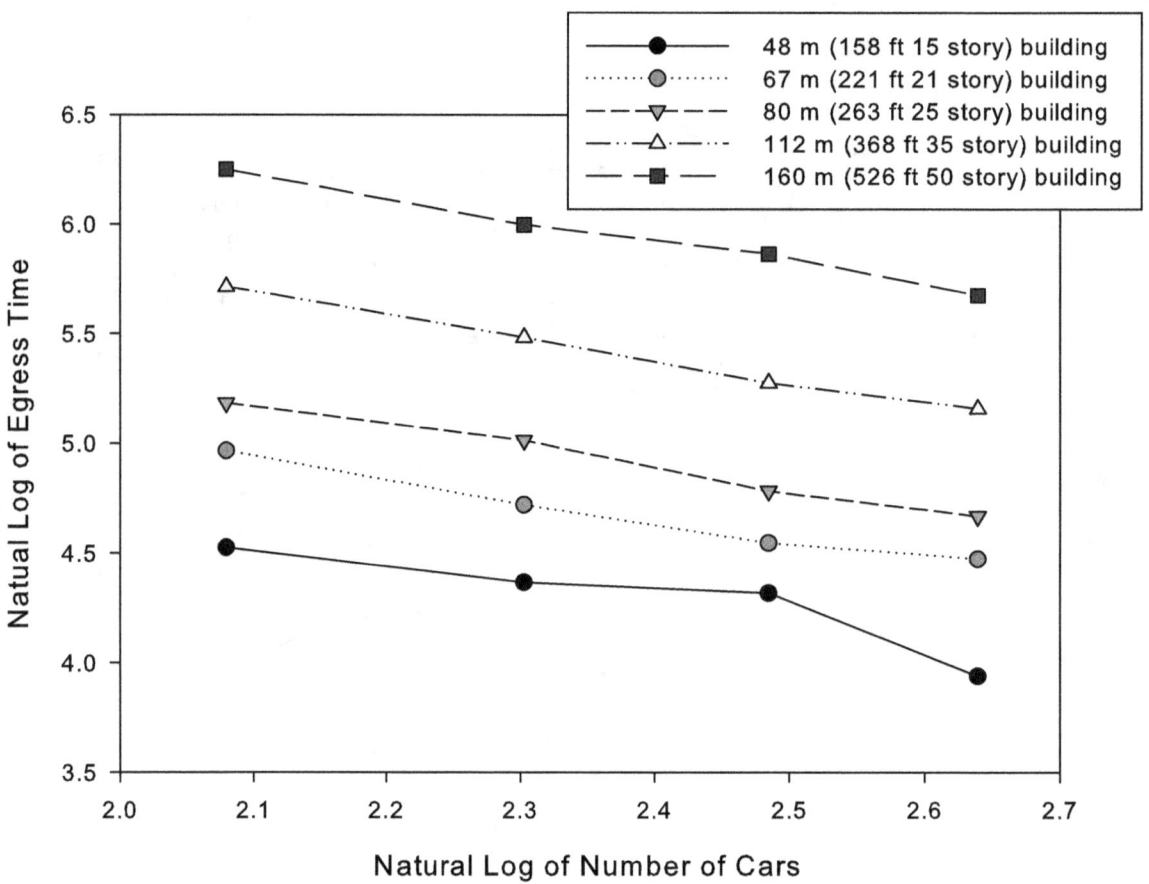

**Figure 12. A log-log plot of the impact of adding cars to the egress time in a number of building designs.**

The plots in Figure 12 show a strong correlation between the model predictions and the inverse relationship given in eq. (14). Table 1 gives the linear least squares fits of the log-log plots.

While the correlation coefficient, $R^2$, is quite high, it declines as the number of floors decreases. The largest deviation from the relationship established in Eq (14) is in the 48 m (157 ft 15 story) case with 12 cars. The deviations arise due to the inefficiency of the elevator control algorithm when the building has many cars and few floors. This inefficiency would be expected in a real building as well. Cars either stand idle, are sent to empty floors, or are underutilized going to floors with small populations.

---

[5] For Eq (12) a log-log plot would be a plot of pairs on a logarithmic axis so if $t_{EE_B,J}$ is the elevator evacuation time calculated by the Egress Estimator for building B with $J$ elevator cars the linear fit of pairs $\left(\log J_i, \log t_{EE_B}\right)$ should be of the form $\log t_{EE_B,J} = C - \log J_i$.

**Table 1. Coefficients and $R^2$ of linear least squares fits from Figure 12**

| Building | Coefficient for log$J$ term | Constant | $R^2$ |
|---|---|---|---|
| 160 m (526 ft 50 story) | -0.994 | 12.404 | 0.992 |
| 112 m (368 ft 35 story) | -1.012 | 11.907 | 0.995 |
| 80 m (263 ft 25 story) | -0.954 | 11.273 | 0.989 |
| 67 m (221 ft 21 story) | -0.896 | 10.900 | 0.977 |
| 48 m (158 ft 15 story) | -0.945 | 10.629 | 0.842 |

### 4.1.4 Elevator Car Capacity

The impact of the capacity of a car is requires using an equation for calculating $t_{r,n}$. The round trip travel time to a floor is the sum of a number of different operations. The equation is

$$t_r = 2t_t + 2t_d + t_l + t_u \tag{13}$$

where $t_t$ is the travel time from the ground floor to floor $r$ or from floor $r$ to the ground floor, $t_d$ is the time it takes to open and close the doors, $t_l$ is the time it takes to load the passengers and $t_u$ is the time it takes to unload the passengers.

Both the times to load and unload are dependent on the number of passengers that are loading. However, they use the same equation with different parameters. The equation is

$$t_{u/l} = \tag{14}$$

$$t_{l/u} = \begin{cases} t_{dw} & N = 1 \\ t_{dw} + t_{in/out}(N-2) & N \geq 2 \end{cases} \tag{16}$$

where $N$ is the number of passengers loading or unloading, $t_{dw}$ is the dwelling time (the minimum time the doors are open) and $t_{in/out}$ is a parameter. The dwelling time is set to 4 s in the Egress Estimator. When loading $t_{in/out} = t_{in} = 1.0$ and when unloading $t_{in/out} = t_{out} = 0.6$. Since the numbers of people loading and unloading are equal, the two versions of Eq (14) can be combined to obtain

$$t_{l+u} = \begin{cases} t_{dw} & N = 1 \\ t_{dw} + (t_{in} + t_{out})(N-2) & N \geq 2 \end{cases} \tag{17}$$

Letting $t_m = t_t + t_d$ and substituting Eq (10), (15) and (17) into (3) yields

23

$$t_e = \frac{1+\eta}{J} N_B \sum_{r-1}^{R} \left[ 2t_{m,r} + 2t_{dw} + (t_{in} + t_{out})(C_c - 2) \right] \tag{18}$$

where $C_c$ is the capacity of the elevator cars and with the assumption that the population on the floor, $P_f$, is an integer multiple of the capacity of the elevator cars and an added assumption that the cars can carry at least two passengers. Note that for a given building and elevator design $T_{m,B} = \sum_{r=1}^{R} t_{m,r}$ is a constant and that $N_B = P_f / C_c$. Substituting into Eq. (18) gives

$$t_e = \frac{(1+\eta)P_f}{JC_c} \left[ 2T_{m,B} + 2Rt_{dw} + R(t_{in} + t_{out})(C_c - 2) \right]$$

$$= \frac{P_{eff}}{J} \left[ \frac{2(T_{m,B} + Rt_{dw})}{C_c} + \frac{R(t_{in} + t_{out})(C_c - 2)}{C_c} \right] \tag{19}$$

$$P_{eff} = (1+\eta)P_f$$

For $C_c = 2$ the time to egress would be.

$$t_e = \frac{P_{eff}}{J}(T_{m,B} + Rt_{dw})$$

As the value of $C_c$ decreases, the first term in Eq. (19) acts like an inverse relationship but as $C_c$ increases $(C_c-2)/C_c$ approaches 1 and so Eq. (19) approaches.

$$t_e = \frac{P_{eff}R}{J}(t_{in} + t_{out})$$

The graph the in Figure 13 shows that the Egress Estimator reproduces the expected shape of the curve. The cases all have 504 people per floor with 8 elevator cars that have a maximum speed of 3.0 m/s and an acceleration of 3.0 m/s². While elevator cars for some of the runs are impractically large they are used to more fully demonstrate the shape of the curves.

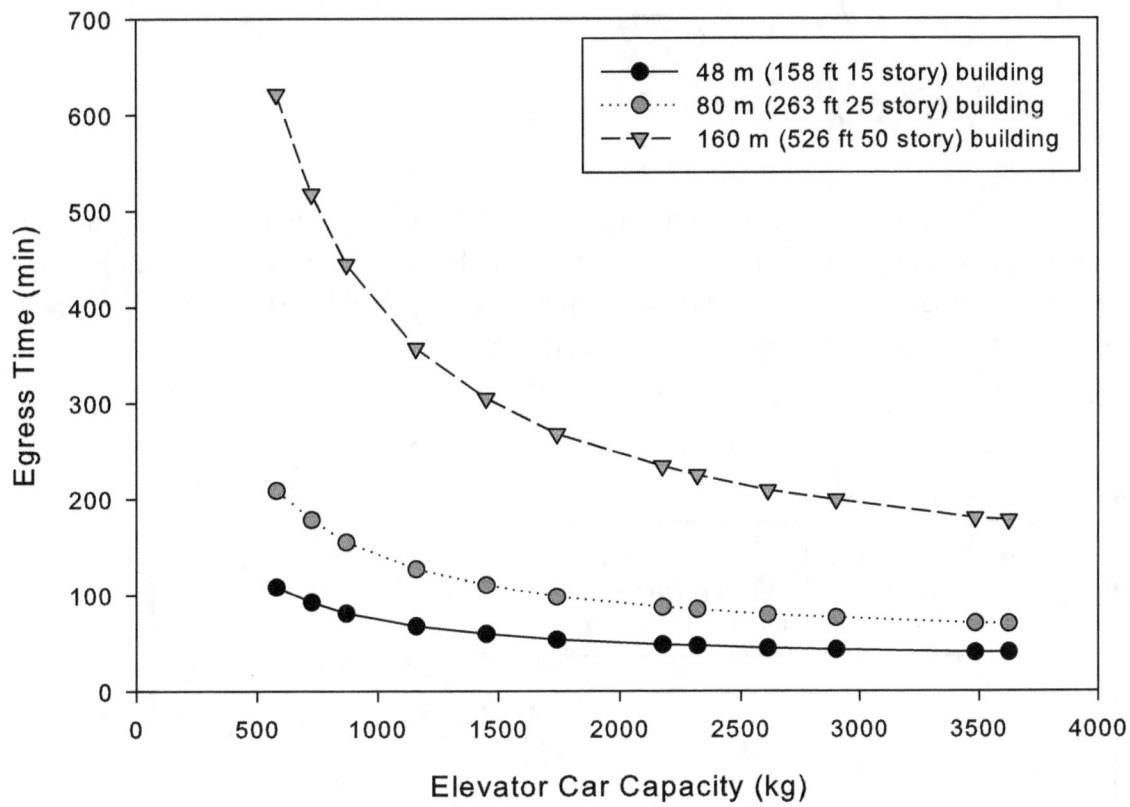

**Figure 13. Impact of the capacity of the elevator cars in three building heights. There are 8 cars per building and 504 people per floor.**

### 4.1.5 Relationship Between Number of Cars and Car Capacity

To look at the relationship between the number of cars, $J$, and the car capacity, $C_c$, it is useful to create a relationship between the two. For example define $C_s = JC_c$, where $C_s$ is the capacity of the elevator system. Substitute $C_s$ into Eq (13) gives

$$t_e = P_{eff}\left[\frac{2\left(T_{m,B} + Rt_{dw}\right)}{JC_c} + \frac{R\left(t_{in} + t_{out}\right)\left(C_c - 2\right)}{JC_c}\right]$$

$$= P_{eff}\left[\frac{2\left(T_{m,B} + Rt_{dw}\right)}{C_s} + \frac{R\left(t_{in} + t_{out}\right)\left(C_c - 2\right)}{C_s}\right]$$

(20)

Eq. (20) can be simplified by grouping constants to

$$t_e = K_m C_c + K_b$$

$$K_m = \frac{P_{eff} R (t_{in} + t_{out})}{C_s}$$

$$K_b = \frac{2P_{eff} \left[ T_{m,B} + R(t_{dw} - 1) \right]}{C_s}$$

(21)

Eq. (21) is linear in $C_c$. Figure 14 shows the nearly linear relationship given in Egress Estimator. The floor population is changed to 384 per floor, so the population is an exact multiple of a wide range of elevator car capacities. As in the last section, calculations are done with elevator cars with impractical capacity to more fully illustrate the curves.

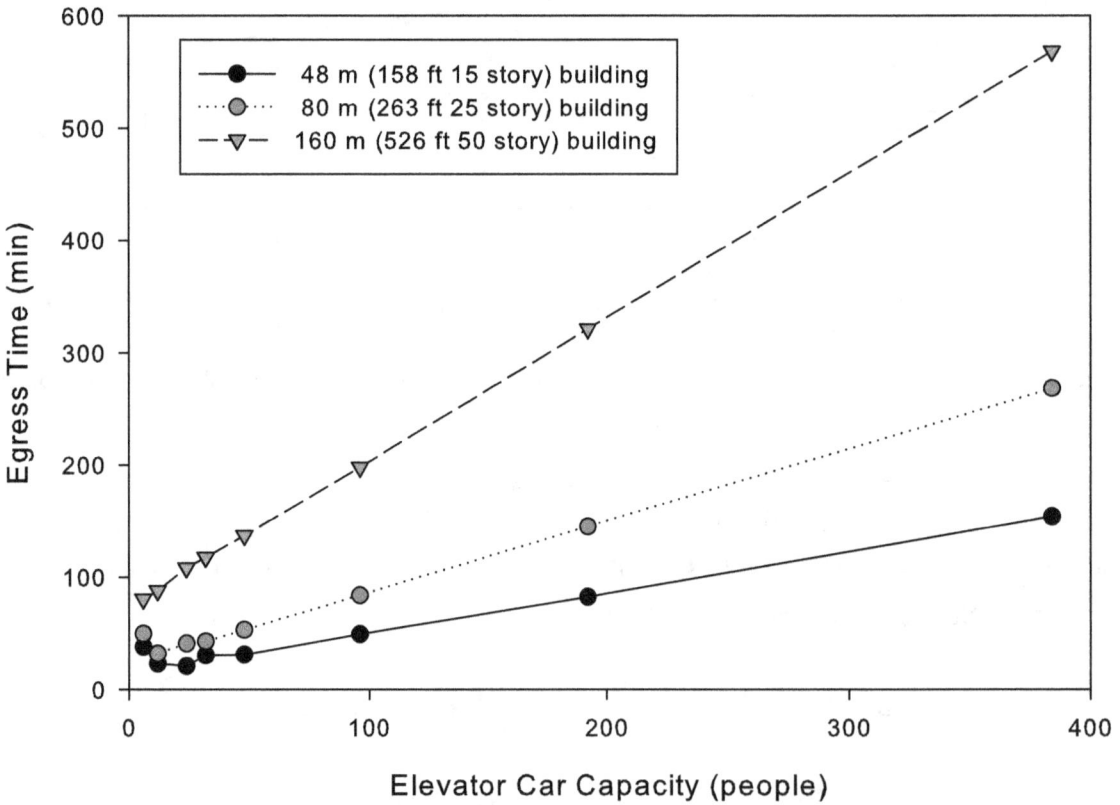

**Figure 14. The impact of increasing elevator car capacity when holding total system capacity constant in three building designs. There are 384 people per floor.**

Notice that in Figure 14, the egress time increases as the car capacity increases, but in Figure 13, the egress time decreases as the car capacity increases, which is the expected result. The reason is that here the capacity of the system doesn't change so the loading and unloading times become significant. In the previous analysis, the number of cars stays constant while the total capacity of the system increases, so egress time falls.

Figure 15 shows a blow up of the part of Figure 14 where the car capacities are less then or equal to 32 people per car. While the data for the 160 m (525 ft 50 story) building is linear, the data for the other two buildings is not. The reason is that a large number of cars are needed to yield the same system capacity when the capacity of each car is relatively small. When a car can only hold 6 people, the system has to have 64 cars to have a capacity of 384 people. In both the 48 m (157 ft 15 story) building and the 80 m (262 ft 25 story) building for the lower capacity cars, there are simply too many cars for the elevator control algorithm to handle efficiently. It seems likely that a significant surplus of cars to floors is not likely to be handled very efficiently; it is also unlikely that any building would be designed with such a surplus.

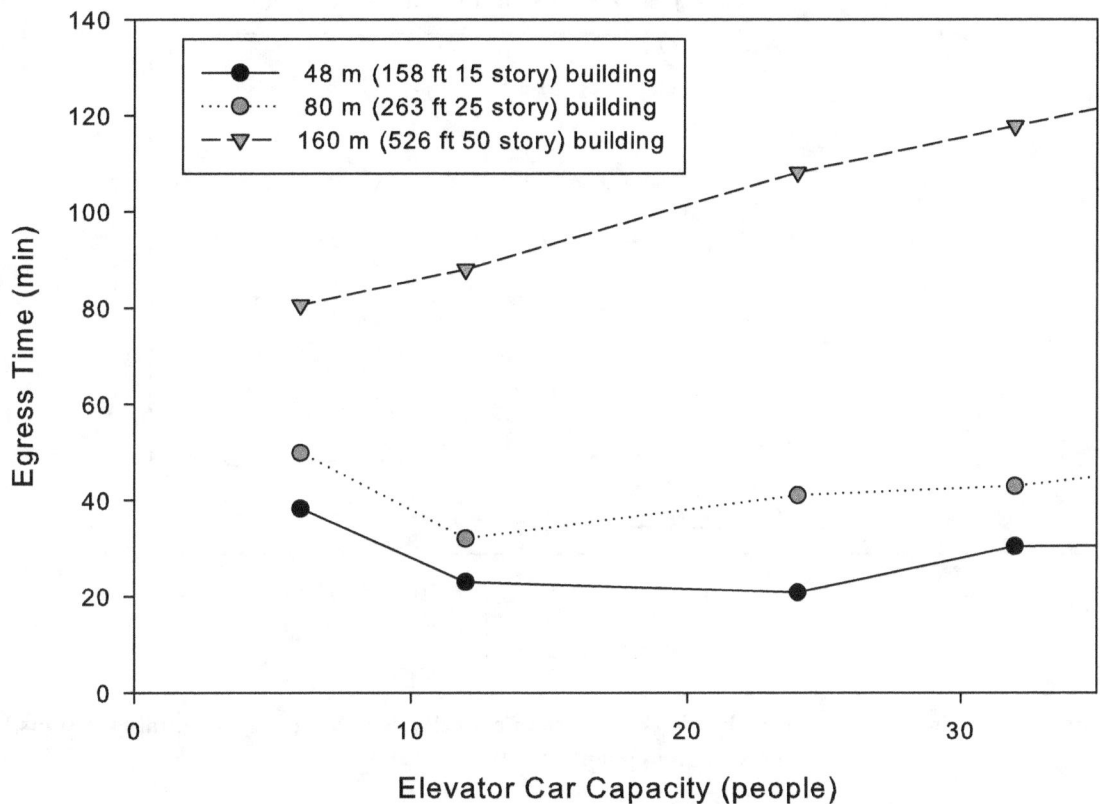

**Figure 15. A detail of the impact of increasing car capacity when holding system capacity constant in three building designs. There are 384 people per floor.**

Since it is assumed that $C_s$ is held constant then $C_c = \frac{C_s}{J}$ . Substituting into Eq (16) gives

$$t_e = P_{eff}\left[\frac{2(T_{m,B} + Rt_{dw})}{C_s} + \frac{R(t_{in} + t_{out})(C_s/J - 2)}{C_s}\right] \qquad (22)$$

Like with Eq. (21), Eq. (22) can be simplified to

27

$$t_e = \frac{K_m C_s}{J} + K_B \tag{23}$$

where $K_m$ and $K_b$ are the same as in Eq. (21). Eq. (23) is of the form of an inverse relationship between the number of cars and the total egress time. Figure 16 shows exactly that kind of a relationship.

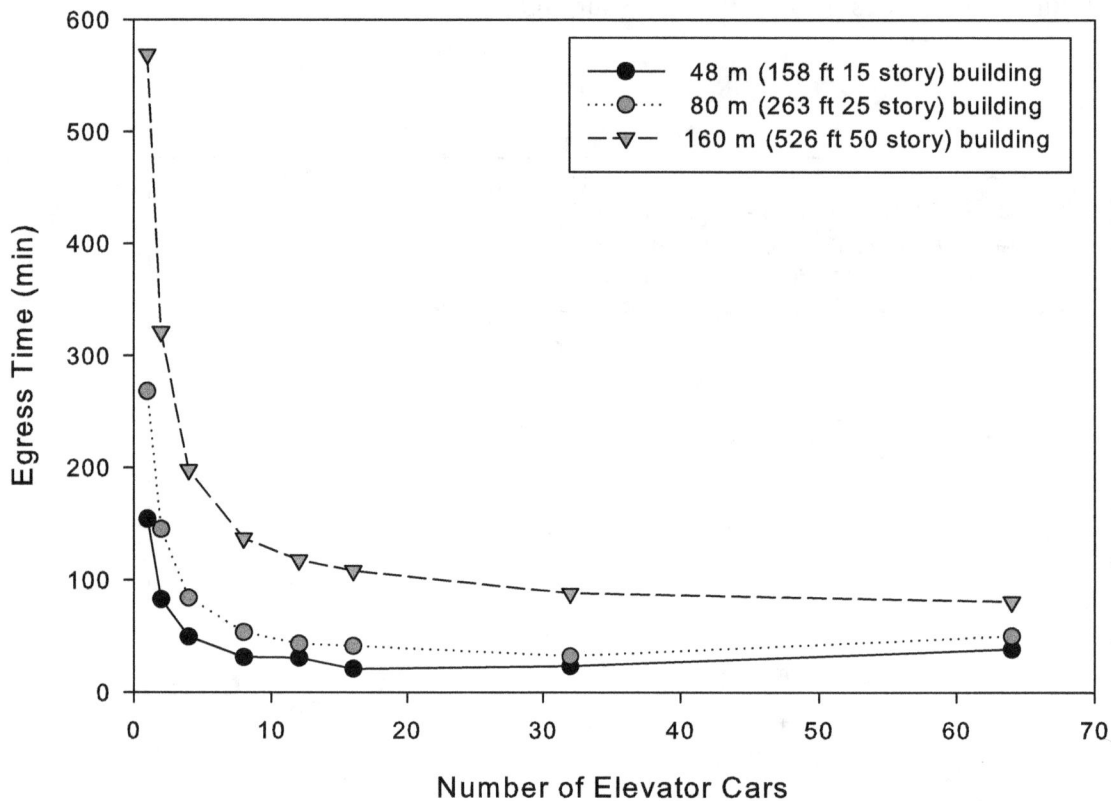

**Figure 16. Impact of adding elevator cars while keeping system capacity constant in three building designs. There are 384 people per floor.**

Note that the inefficiency evident in Figure 15 is not as visible when plotting results against the number of cars. However, on closer examination, the 64 car cases for both the 48 m (157 ft 15 story) and 80 m (262 ft 25 story) buildings are slower than the 32 car cases.

### 4.1.6 Maximum Velocity and Acceleration

The last aspect of an elevator design considered is the impact on egress time of the mechanical parts of the elevator system. For this analysis, the number of cars and the capacity of the cars is held constant so the capacity of the system is constant. This also means that

$T_{L/U} = \dfrac{R(t_{in} + t_{out})(C_s - 2)}{C_s}$ can be defined and substituted into Eq (19) to yield

28

$$t_e = \frac{(1+\eta)P_f}{C_s}\left[2\left(T_{m,B} + t_{dw}R\right) + T_{L/U}\right] \qquad (24)$$

Since the term $t_{dw}R$ is small compared to $T_{m,B}$ and even $T_{L/U}$ and noting that its effect on total evacuation time is linear, it will not be considered further. The details of how the travel time from one floor to another is calculated are covered in reference [11] and will not be discussed here. It is sufficient to say for the purposes here that the impact of changing the maximum travel speed and the acceleration are nonlinear but act in the expected way. In other words, increasing the maximum speed or acceleration will reduce the egress time. Furthermore, there are a large number of design considerations that go into specifying an elevator system so that it is unlikely that the egress time will be a major criterion.

However, it is certainly useful to observe the impact of changing both the elevators maximum speed and acceleration. Figure 17 shows the impact of increasing the maximum velocity from 3.0 m/s to 11 m/s for 48 m (157 ft 15 story), 80 m (262 ft 25 story) and 160 m (525 ft 50 story) buildings. Each building has only two people per floor and one elevator car to maximize the impact of the differences in speed. Because of the significant differences in time scales for the three buildings each curve was normalized by the time it took to empty the building with a single car with maximum velocity of 3.0 m/s.

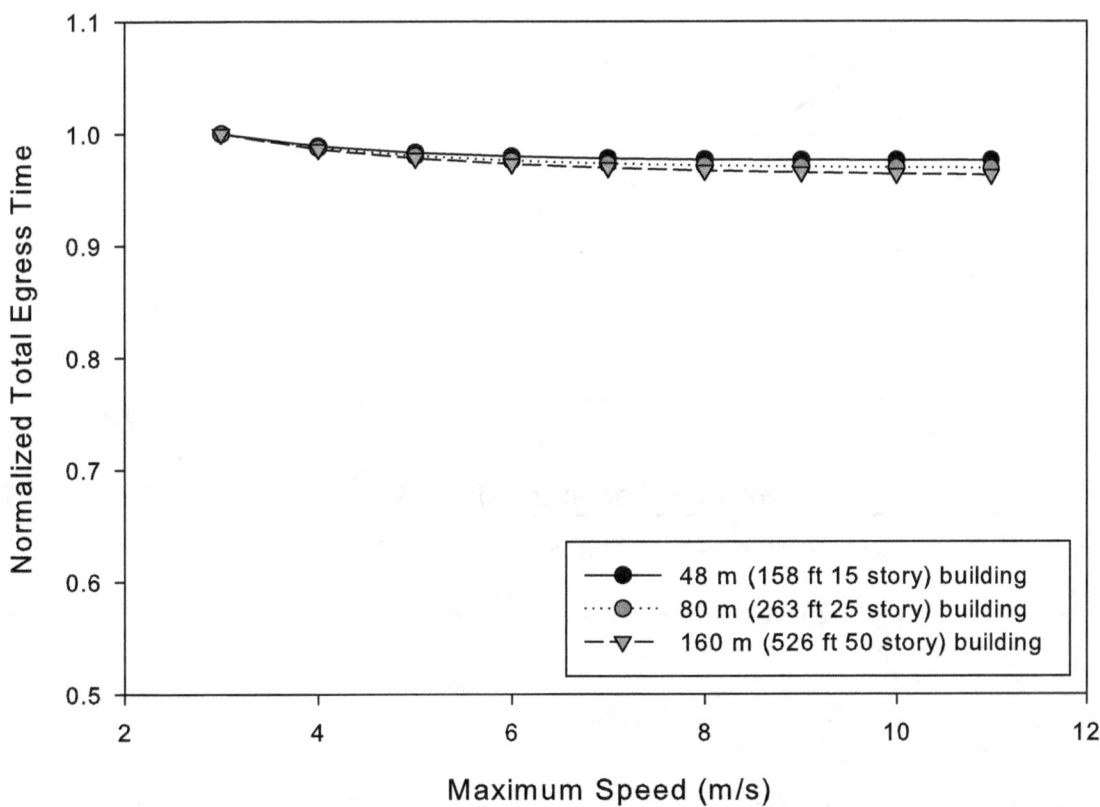

Figure 17 Relative impact of increasing the maximum velocity of the elevator cars in three building designs.

Even for the 160 m (525 ft 50 story) building the improvement in egress time is less than 4 % since for many of the lower level floors, the car does not reach maximum velocity. In addition, at a higher maximum velocity, it takes a longer time to reach a maximum so the distance travelled at maximum velocity is shorter for higher velocities.

The story changes if just the top 5 floors of each building are considered. Figure 18 shows the impact in the change of maximum velocity for just the top 5 floors of a 48 m ( 157 ft 15 story), 80 m (262 ft 25 story) and 160 m (525 ft 50 story) buildings. In the 160 m (525 ft 50 story) building, the top five floors evacuate more than 38 % faster. More discussion of the impacts of zoned evacuation are in section 4.2.

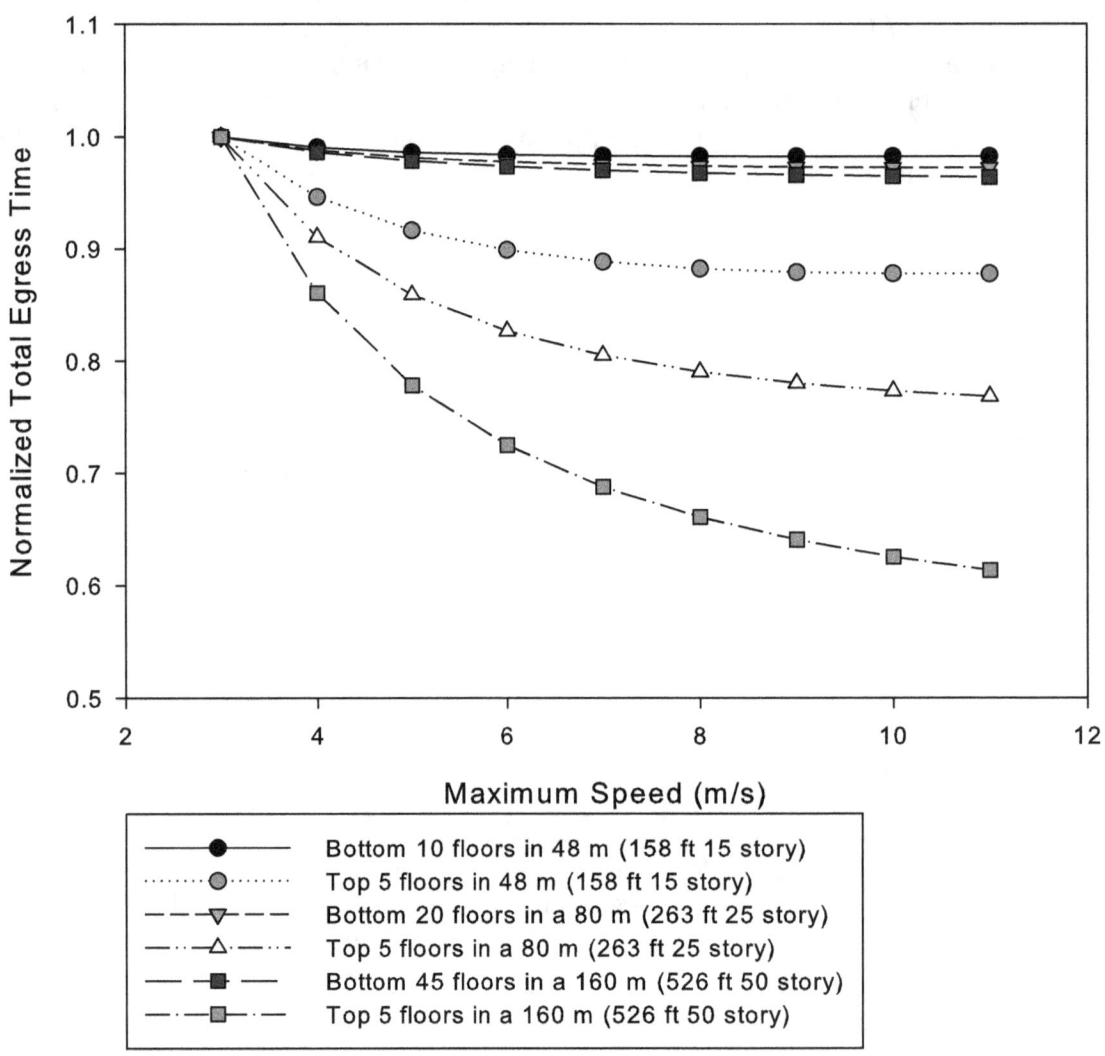

| | |
|---|---|
| ——●—— | Bottom 10 floors in 48 m (158 ft 15 story) |
| ········○········ | Top 5 floors in 48 m (158 ft 15 story) |
| — —▽— — | Bottom 20 floors in a 80 m (263 ft 25 story) |
| —··—△—··— | Top 5 floors in a 80 m (263 ft 25 story) |
| — ■ — | Bottom 45 floors in a 160 m (526 ft 50 story) |
| —·—■—·— | Top 5 floors in a 160 m (526 ft 50 story) |

**Figure 18 Relative impact on egress time of increasing maximum velocity for the top 5 floors for three building heights.**

30

While exploring the impact of changing acceleration it is important to remember that changes in acceleration will put an added stress on the occupants so the window of possible accelerations are limited. Still, it is illustrative to know the impact of relatively small changes in acceleration have on the egress time. To examine the impact of changing acceleration start with a 15 story building with three different maximum velocities, 3.0 m/s, 7.0 m/s and 11.0 m/s.

The motivation for looking at just the 48 m (157 ft, 15 story) building is to consider the impact of increased acceleration. Suppose at the initial value for acceleration, the elevator car reaches full speed by floor $R_0$ so that the car reaches maximum speed before slowing down to a full stop traveling to floor $2R_0$, say $t_{a0}$. Increasing the acceleration will reduce the amount of time the car takes to get from the ground floor to $R_0$ and so reduce the time it takes to get to $2R_0$, to $t_{a1} < t_{a0}$. However, for any floor above $2R_0$ the time to reach the floor will be $t_{a0}$ or $t_{a1} + nt_f$ where n is the number of floors about $2R_0$ and $t_f$ is the time it takes the elevator car to travel one floor. So the impact of increasing acceleration should be most visible in shorter buildings.

First note the small relative change between the curve for 7.0 m/s and 11.0 m/s and remember from Figure 17 that the changes in total egress time decreases as the speed increases. Second, notice the very small improvement in speed as acceleration increases. The total change in egress time for varying acceleration from 3.0 m/s$^2$ to 3.8 m/s$^2$ is less than 50 s for all three speeds.

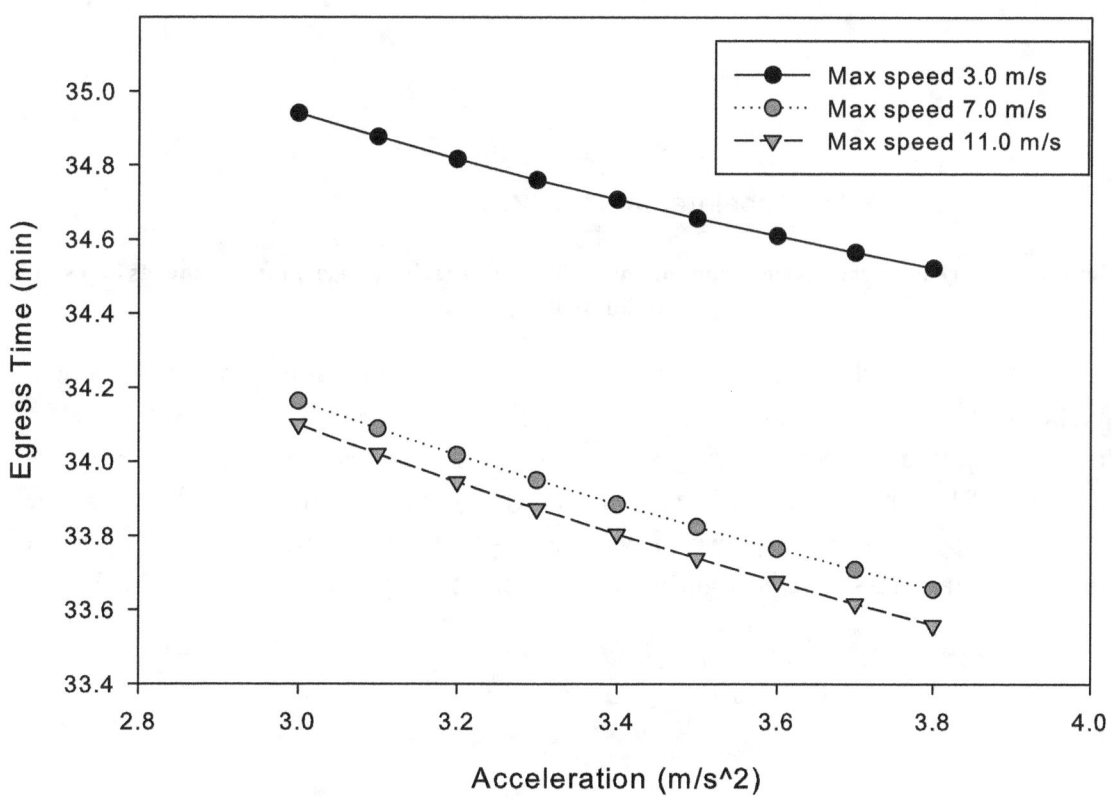

**Figure 19. Impact of increasing the accleration for three different maximum elevator car speeds in a single 15 story building design.**

31

To check that the analysis of impact of increased acceleration on higher buildings see the results are the same for a 80 m (262 ft 25 story) building as shown in Figure 20. Improvements in acceleration translate into very small improvements in egress time.

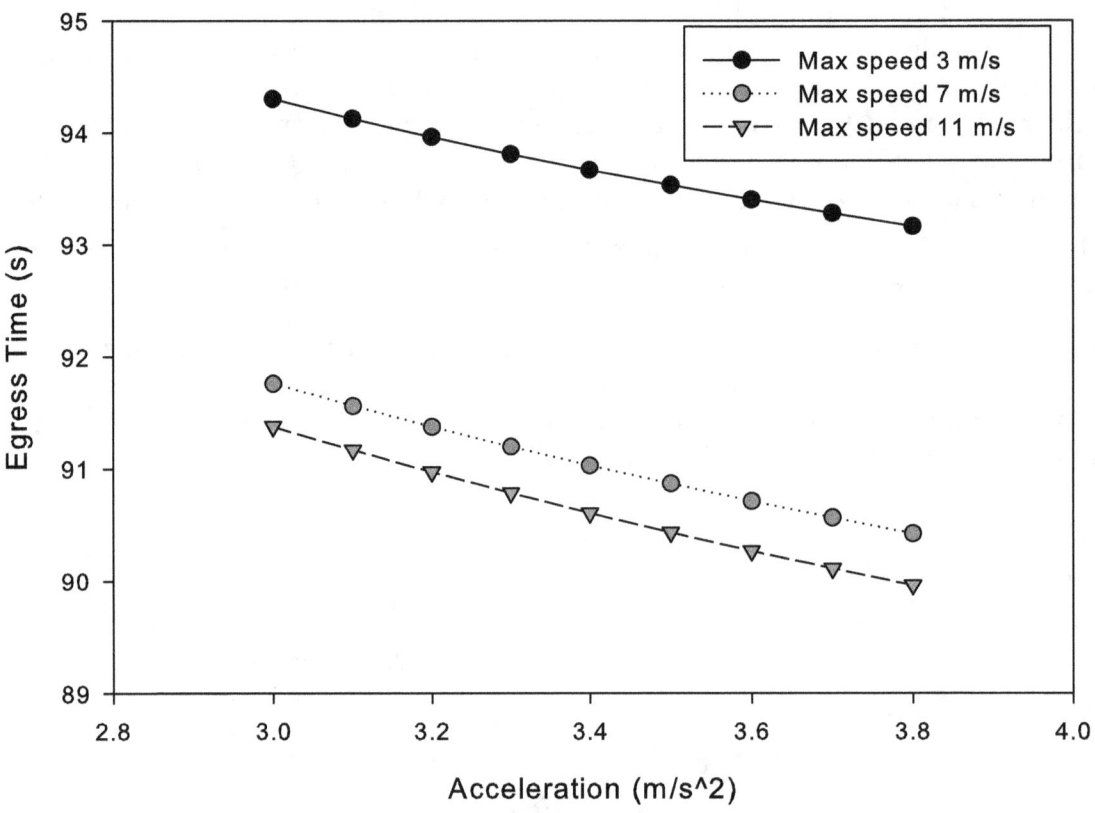

**Figure 20. Impact of increasing the acceleration for three different maximum elevator car speeds in a single 25 story building design.**

In order to examine the impact of the change in acceleration, the acceleration and egress time for both buildings can be normalized. Figure 21 shows that the same speed in the 15 story and 25 story buildings have similar and small changes with respect to changes in acceleration. For a maximum velocity of 11.0 m/s, a 27 % increase in acceleration only decreases total egress time by less than 3 %. The systems are very insensitive to changes in acceleration. Also note that the relative improvement in egress time is slightly larger in the 15 story building then in the 25 story building.

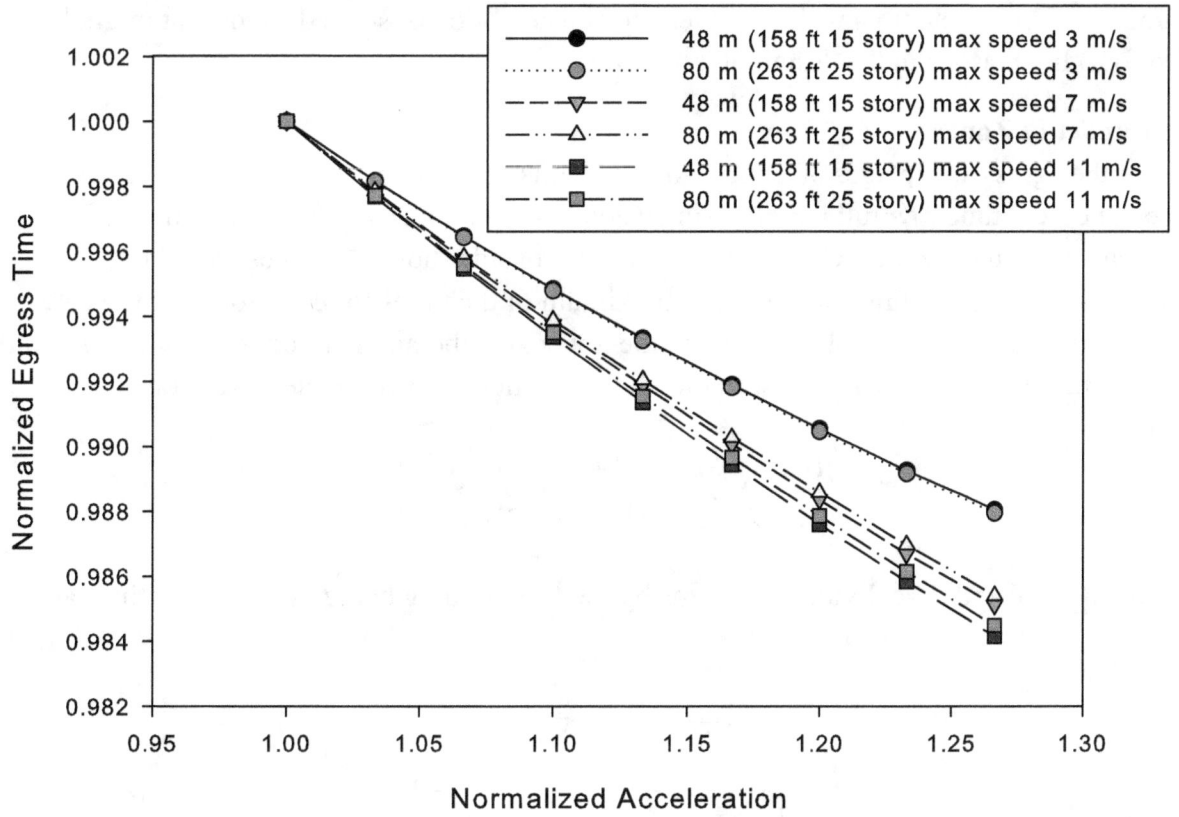

**Figure 21. Normalized estimated egress time as a function of normalized acceleration for several building heights and maximum elevator velocities.**

These same small improvements in egress time are seen with changes in where the cars stop during an evacuation. As an example, consider a 15 story building scenario:

- 48 m (158 ft, 15 story) building

- 66 occupants per floor

- 6 elevators, 12 person capacity, 3 m/s maximum speed.

Note that the total population per floor is set so that there would be a partial load on the last trip from any floor. Total evacuation time this scenario is estimated to be 915 s if the elvators stop at additional floors when occupant load on the elevators is less than 80 % (consistent with the elevator evacuation algorithm presented in section 2.3.2). If the elevators do not stop for additional passengers (consistent with the current requirements of ASME A17.1), the total evacuation time increases, but by less than 5 %, to 940 s.

## 4.2 Zoned Elevator Evacuation

There are a number of reasons a building design might include elevator zones, including handling the normal times of high elevator use and maximizing leasable space. The cost and benefits of such concerns are outside the scope of this study. The focus here is only on the impact of zones on evacuation, which will lead to consideration of some designs that might be rejected for other reasons in the design of a building.

### 4.2.1 Impact of Zones

To look at the egress advantages of using two zones, it is useful to have a baseline. One baseline is to use the egress time of a full building single zone evacuation using the total number of elevator cars used in the zones. Going back to Eq. (3) consider how two zones would be analyzed. The total egress time would be the maximum of the times for each zone. So the egress time for two zones with the number of cars in the lower zone being the fraction $0.0 < f_e < 1.0$ and the upper zone have the other $1-f_e$ cars. Let $R_l$ be the top floor served by the lower zone.

$$t_e = \max\left(\frac{1+\eta}{f_l J}\sum_{r=1}^{R_1}\sum_{k=1}^{T_r}t_{r,k}, \frac{1+\eta}{(1-f_l)J}\sum_{r=R_1+1}^{R}\sum_{k=1}^{T_r}t_{r,k}\right) \qquad (25)$$

To find the normalized time divide both terms by the full building being evacuated with $J$ cars to get

$$t_{e,norm} = \max\left(\frac{\sum_{r=1}^{R_1}\sum_{k=1}^{T_r}t_{r,k}}{f_l\sum_{r=1}^{R}\sum_{k=1}^{T_r}t_{r,k}}, \frac{\sum_{r=R_1}^{R_B}\sum_{k=1}^{T_r}t_{r,k}}{(1-f_l)\sum_{r=1}^{R}\sum_{k=1}^{T_r}t_{r,k}}\right) \qquad (26)$$

Recognize that the numerators of the two terms must add up to the denominator and the equation reduces to

$$t_{e,norm} = \max\left(\frac{f_{lz}(R_l)}{f_e}T, \frac{(1-f_{lz}(R_l))}{(1-f_e)}T\right)$$

$$T = \sum_{r=1}^{R}\sum_{k=1}^{T_r}t_{r,k} \qquad (27)$$

where $0.0 < f_{lz}(R_l) < 1.0$ is a function that returns the fraction of the total egress time $T$ the first $R_l$ floors take to evacuate. It is important to note that the fraction of $T$ is the fraction of time and not the fraction of floors. As discussed in section 4.1.2 there is a quadratic relationship between the number of floors and the egress time. Figure 22 shows the function $f_{lz}(r)$ in a 48 m (157 ft 15 story) building.

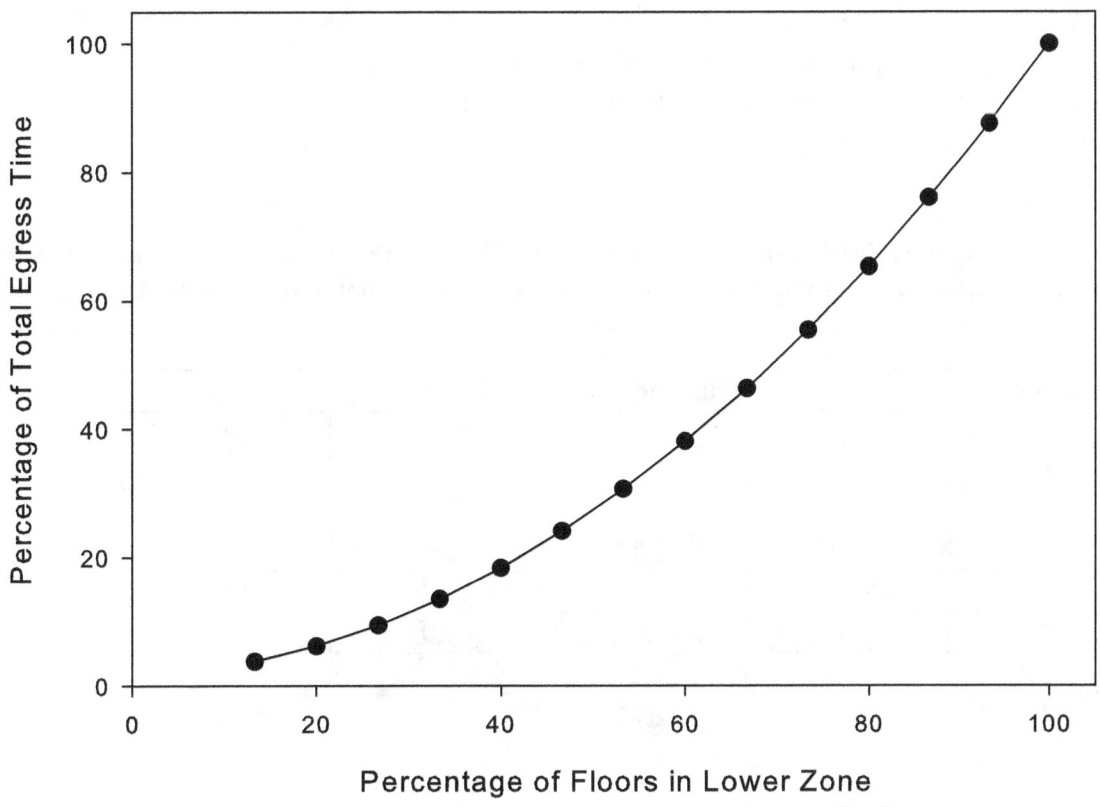

**Figure 22 Percentage of Floors in 48 m (157 ft 15 story) building in the lower zone vs the percentage of the total egress time of the building**

From Eq. (23) Table 2 can be generated for a 48 m (157 ft 15 story) building with 8 elevator cars. It shows that having zones increases the egress time.

**Table 2. Total egress time from Eq. (27) for a 48 m (157 ft 15 story) building with 8 elevator cars divided into two zones normalized by the total egress time for a single zone with 8 cars. Maximum elevator speed is 3 m/s in both zones.**

| Number of Floors in lower zone | Number of cars in lower zone | | | |
|---|---|---|---|---|
| | 3 | 4 | 5 | 6 |
| 7 | 1.24 | 1.07 | 1.42 | 2.13 |
| 8 | 1.42 | 1.07 | 1.24 | 1.87 |
| 9 | 1.60 | 1.20 | 1.07 | 1.60 |
| 10 | 1.78 | 1.33 | 1.07 | 1.33 |
| 11 | 1.96 | 1.47 | 1.17 | 1.07 |
| 12 | 2.13 | 1.60 | 1.28 | 1.07 |

Table 3 shows the calculated values using the Egress Estimator. The data are normalized egress time for 504 people per floor in a 48 m (157 ft 15 story) building. There are 8 cars in total each able to carry a maximum of 8 passengers with the maximum speed of cars in both zones 3.0 m/s. The times are all normalized by the time it would take a single zone of 8 cars that have a maximum speed of 3.0 m/s. The cells shaded yellow have a normalized time < 1.0. The cells shaded green have normalized times greater than 1.0 but less than 1.25, the cells shaded blue are between 1.25 and 1.5 and the rest are > 1.5.

**Table 3. Total egress time from Egress Estimator for a 15 story building with 8 elevator cars divided into two zones normalized by the total egress time for a single zone with 8 cars. Maximum elevator speed is 3 m/s in both zones.**

| Number of floors in lower zone | Number of cars in lower zone | | | |
|---|---|---|---|---|
| | 3 | 4 | 5 | 6 |
| 7 | 0.98 | 1.22 | 1.63 | 2.44 |
| 8 | 1.19 | 1.14 | 1.57 | 2.20 |
| 9 | 1.30 | 0.97 | 1.26 | 1.89 |
| 10 | 1.49 | 1.24 | 1.08 | 1.66 |
| 11 | 1.78 | 1.30 | 1.02 | 1.30 |
| 12 | 1.90 | 1.46 | 1.28 | 1.05 |

There are some significant differences between Table 2 and Table 3. Two of the configurations have a normalized time less than 1.0 in Table 3. The basic form with the diagonal having the lowest times with a rough symmetry about the diagonal holds but individual entries vary significantly. The most likely explanation given the analysis with Eq. (3) is that the efficiency is improved with 4 cars serving fewer floors than with 8 cars serving all the floors. This analysis gives another reason for using a control algorithm instead of just assuming an single constant factor to account for inefficiency.

### 4.2.2 Impact of Car Speed

Table 4 is structured in the same way as Table 3 except the maximum number of passengers in a car are 10 and the cars going to the upper zone travel 11.0 m/s maximum, while the cars in the lower zone have a maximum speed to 3.0 m/s

Table 4. Total egress time from Egress Estimator for a 48 m (157 ft 15 story) building with 8 elevator car normalized by the total egress time for a single zone with 8 cars. Maximum elevator speed is 11 m/s in the upper zone and 3 m/s in the lower zone.

| Number of floors in lower zone | Number of cars in lower zone | | | |
|---|---|---|---|---|
| | 3 | 4 | 5 | 6 |
| 6 | 0.93 | 1.29 | 1.50 | 2.29 |
| 7 | 0.96 | 1.01 | 1.35 | 2.01 |
| 8 | 1.23 | 0.94 | 1.31 | 1.79 |
| 9 | 1.31 | 0.98 | 1.02 | 1.53 |
| 10 | 1.49 | 1.25 | 0.92 | 1.31 |
| 11 | 1.78 | 1.31 | 1.02 | 1.03 |
| 12 | 1.90 | 1.46 | 1.29 | 0.96 |

The previous analysis is looking at the difference between using one zone and using two zones. One of the key assumptions is that the elevator cars have the same maximum speed and acceleration in both cases. Having faster cars in one of the zones is not amenable to any extra analysis using eq. (3). The advantages of faster cars only appear when the zone that has the faster cars takes longer than the other zone. Typically, because of trip distance, the upper zone is the one that takes the longest, but this is not necessarily the case.

## 4.3 Phased Elevator Evacuation

The phased evacuation follows the same scheme as in section 3.2. Only the fire floor, the two floors above and the two floors beneath are evacuated. The population is evacuated to the floor 3 floors beneath the fire floor. In the Egress Estimator, this is modeled as the evacuation of a 6 story building. To look at the impact of the elevators on phased evacuation, a number of scenarios were run with different numbers of cars and capacities of cars. The base scenario assumes that cars are called to the safe refuge floor, which is three floors down from the evacuation floor. Then 5 floors are evacuated, the two above, the evacuation floor and the two below. The data are shown graphically in Figure 23.

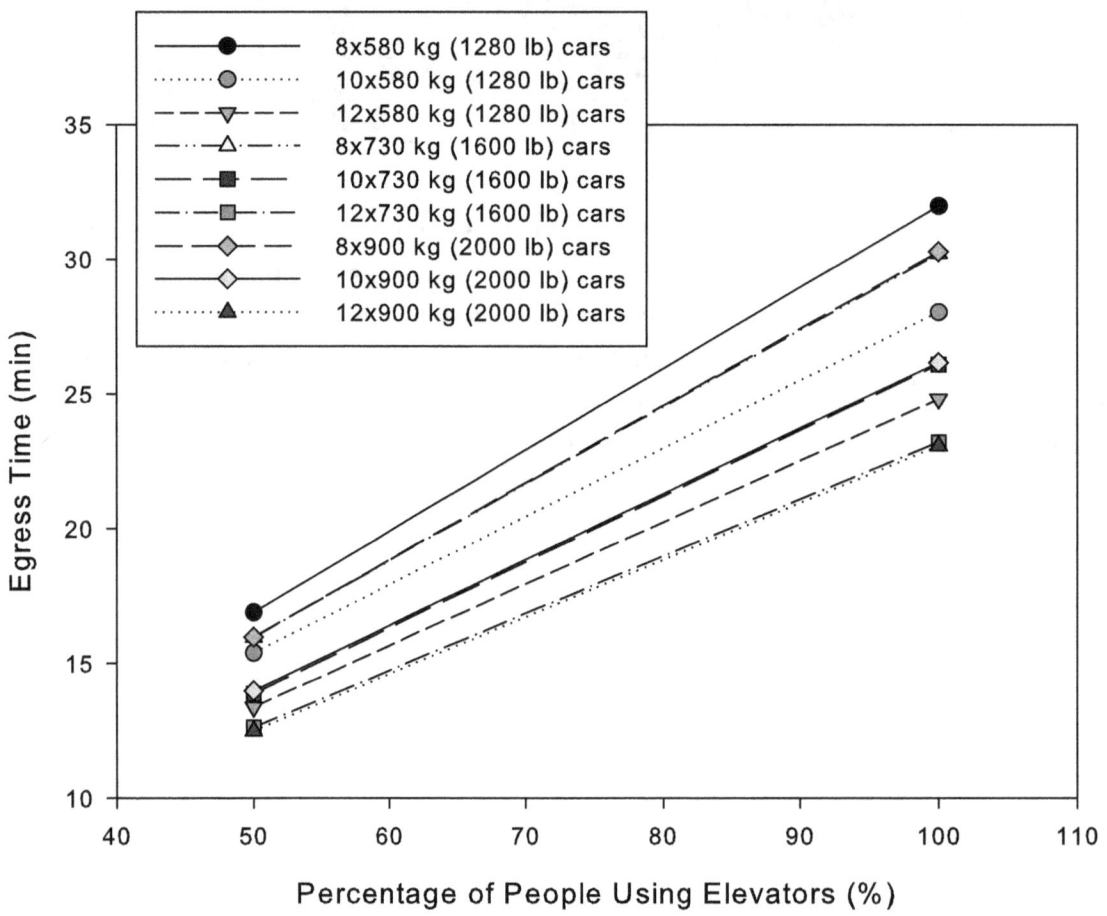

Figure 23. Impact on estimated egress time for phased elevator egress for several elevator configurations.

The graph shows the expected results of reducing the number of people using the elevators reduces the egress time. Increasing the number of cars or the car capacity also decreases egress time. However, Figure 24 shows that there is little improvement between using 10 cars and 12 cars for each capacity. In section 4.1.3, it was shown that total egress time has an inverse relationship with number of elevator cars, which means every new car added reduces total egress time by a smaller amount than the previous car.

38

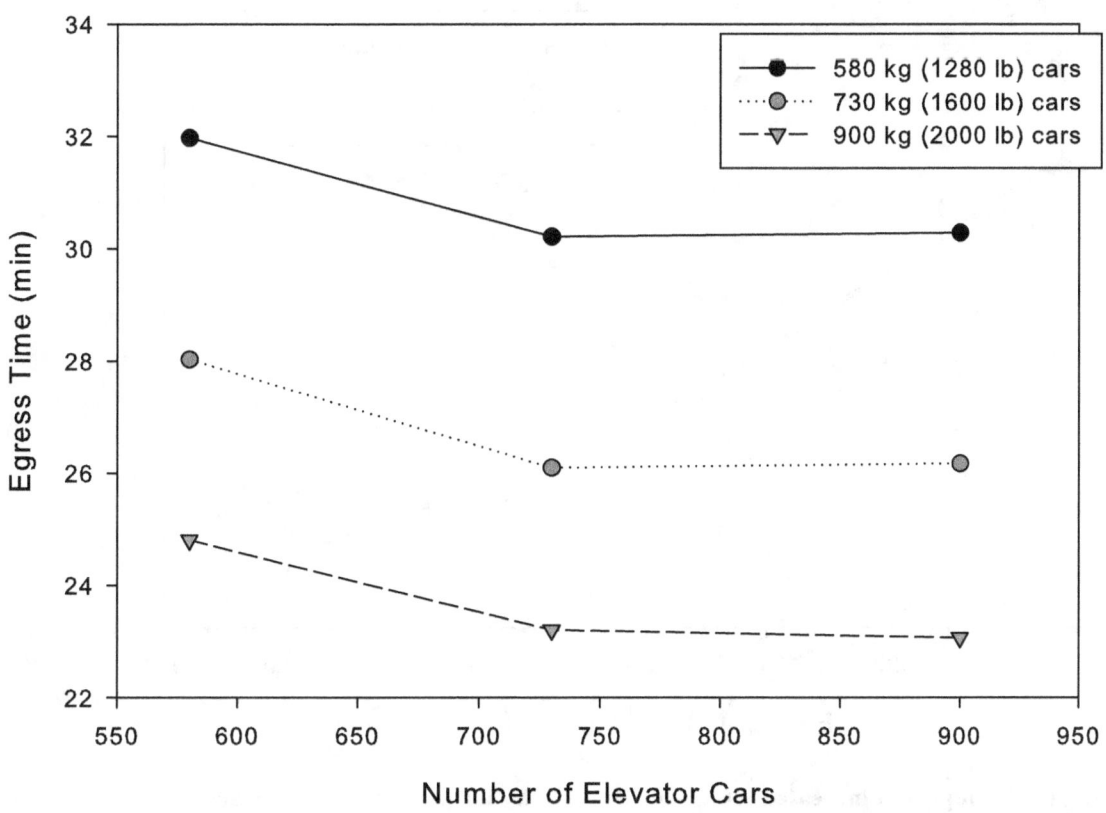

**Figure 24. Impact of increasing the number of elevator cars in a phased evacuation using only elevators. Total elevator capacity is held constant.**

In fact, for the 8 person capacity cars and the 10 person capacity cars the egress time increases only about 4 s and 5 s respectively. The inefficiency of the extra cars means there is no real improvement to be had by increasing the capacity of the system with more cars. Increasing the capacity of the system by increasing the capacity of the cars is sometimes effective as is seen in Figure 25. An increase in car capacity from 8 people to 12 people decreases the egress time about 23 %. As observed before, going from 10 cars to 12 cars doesn't reduce the total egress time.

As discussed in section 4 the Egress Estimator does not send an elevator car to a floor that already has a car serving it. Considering that increasing the capacity of the elevator cars does decrease the total egress time, it is reasonable to think that if the rules were changed such that two cars could serve a floor at a time, the egress time would decrease similarly to doubling the capacity of the elevator cars. Currently, the rules governing elevator cars cannot be changed in the Egress Estimator.

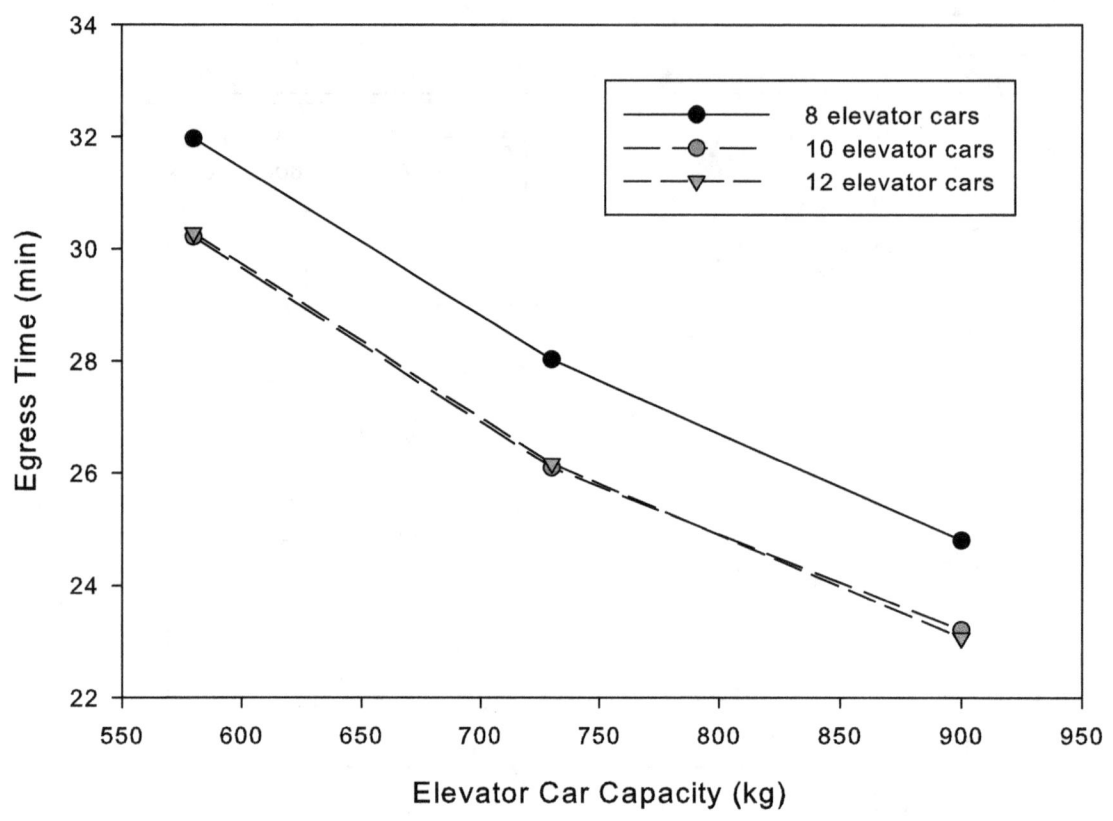

**Figure 25 Impact of increased total elevator capacity for phased elevator evacuation.**

# 5 Combined Stairwell and Elevator Evacuation

In this section, a general method will be presented to take the calculations from the Egress Estimator and develop an analysis of a particular egress design. The first section will look at the method. The second section will look at the types of uncertainty that have to be considered.

## 5.1 Combining Stairs and Elevators

As noted previously in sections 3 and 4 the egress time for both stairs and elevators are linear with respect to the number of people using each system. This suggests a method of analysis of a building evacuation system.

For reasons that will be explained momentarily first note that for either stairs or elevators, the amount of time it takes to evacuate is 0 s if no one is using the system. While the lines formed by calculations of the Egress Estimator do not actually go through the 0 value for either the stairs or the elevator, they are so close that the values serve as a good approximation.

For a particular egress system in a particular building, one can make a run of the Egress Estimator with some fraction of the population using the elevators and the rest using the stairs, say 50 %. The Egress Estimator provides an estimate of the total egress time and it provides the time each system takes to evacuate. Assume these times are ($t_{el}$, $t_{st}$).

Consider an independent variable, $f_{el}$, representing the fraction of people using the elevator. When $f_{el} = 0.0$, no one is using the elevator so the time to complete the elevator evacuation is 0.0 s. When $f_{el} = 1.0$ the number of people using the stairs is 0.0, so the evacuation time for the stairs is 0.0 s. These assumptions give two sets of points that define two lines, {(0.0, 0.0), (0.5, $t_{el}$)} for the elevators and {(0.5, $t_{st}$), (1.0, 0.0)} which defines two lines as follows

$$t_{el} = 2t_{el,0.5}f_{el}$$
$$t_{st} = 2t_{st,0.5}\left(1 - f_{el}\right)$$

(24)

The fraction of people using the elevator that minimizes escape time can be found when $t_{el} = t_{st}$ because total egress time is $t_e = \max\left(t_{el}, t_{st}\right)$. The value of $f_{el}$ occurs at

$$2t_{el,0.5}f_{el} = 2t_{st,0.5}\left(1 - f_{el}\right)$$
$$f_{el} = \frac{t_{st,0.5}}{t_{st,05} - t_{el,0.5}}$$

(29)

The minimum time it takes to evacuate the building when the fraction taking the elevator is ideal is

$$t_{e,min} = \frac{2t_{el,0.5}t_{st,0.5}}{t_{st,0.5} + t_{el,0.5}}$$

(30)

Figure 26 graphically shows the intermediate result. Using the new value for $f_{el}$, a second calculation is made with the Egress Estimator. With these second points, a new set of lines can be calculated and the minimum determined as before.

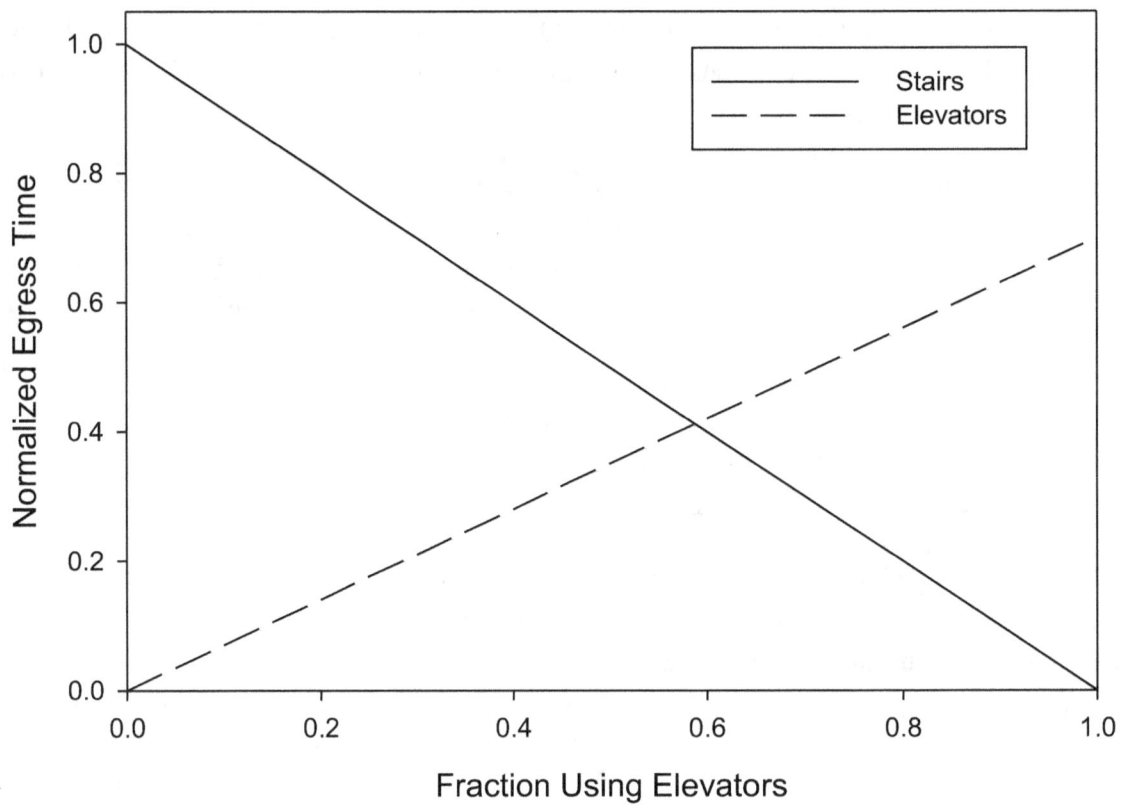

**Figure 26 Initial analysis of ideal fraction using the elevator**

As an example, take a 20 story office building with 120 people per floor given as an example in *The Vertical Transportation Handbook* [16]. It is modified somewhat to keep that analysis simple. In the Handbook, the upper zone elevators let people out on the 10th story, but for the Egress Estimator the upper zone elevators let people out on the ground floor.

The egress system has 2 stairs, each 1118 m (44 in) wide. There are two banks of elevators each with 6 cars with a capacity of 19 people. The maximum speed for the bank that serves floor 1 to 11 is 2.50 m/s while the bank that serves floors 12 to 20 has a maximum speed of 3.50 m/s. A single run of the Egress Estimator with 50 % people on each floor using the elevators results in 12.9 min to complete the stair evacuation while the lower bank of elevators takes 8.3 min and the upper bank takes 9.2 min.

Applying Eq. (25) and (26) gives an ideal faction equal to 0.585 of the population using the elevators and an egress time of 10.8 min total egress time. Figure 27 shows the design curve.

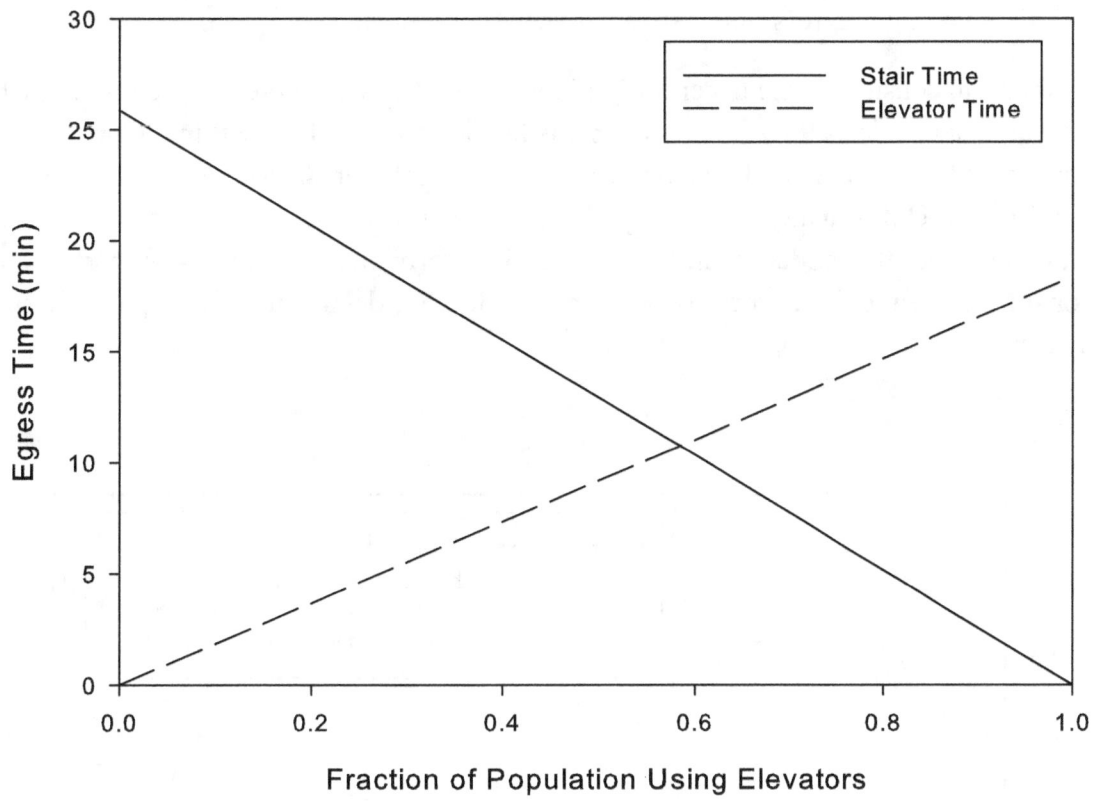

**Figure 27 Analysis of the ideal fraction in a building system**

The next step is to check the results by making a second run with the Egress Estimator using the ideal fraction of 0.585. The results are a total evacuation time of 11.0 min. The lower bank of elevators finish in 9.10 min and the upper bank take 10.2 min. The results are not exact, partly since this is a two zone example but primarily because individual people cannot be divided into fractions of people so they cannot be divided up as exactly as the ideal fraction would suggest.

## 5.2 Uncertainty

As part of the analysis, the uncertainty of the calculation needs to be considered. First consider the uncertainty of the fraction of people using the elevator. While the ideal fraction using the elevator is 0.585 (or rounded up to 0.6), it is difficult to assure that exactly 6 out of every 10 people use the elevator and the rest use the stairs.

Suppose there is reason to believe that the fraction of people using the elevator is between 0.5 and 0.7. What is the impact on the evacuation time? Since the minimum time is when the fraction is 0.6 it is clear the effect will be longer evacuations. Looking at Figure 27 it is clear that if a smaller fraction uses the elevator than the stairs will be the slower component of the system. The egress time can be determined by just calculating the stair egress design line in Eq. (26), which gives 12.9 min, which was calculated earlier. If the fraction using the elevator is 0.7, than the elevator component will be the slower part and the egress time can be found with the elevator design line in Eq. (22), which gives 11.8 min. If the fraction using the elevators varies between 0.5 and 0.7 the maximum egress time varies between 11.0 min and 12.9 min.

One uncertainty to consider is the uncertainty of the calculations. As noted in section 2.1, when compared to the data at least 98 % of all occupants had left the building within the predicted evacuation time plus 30 %. So adding a design margin of 30 % for stairs might give a better performance curve. Unfortunately there is no data to compare with the elevator model. Since with elevators most of the model predicts mechanical performance and not human performance, it is reasonable to assume there would be less uncertainty. For demonstration purposes, assume a design margin of 10 % of predicted time.

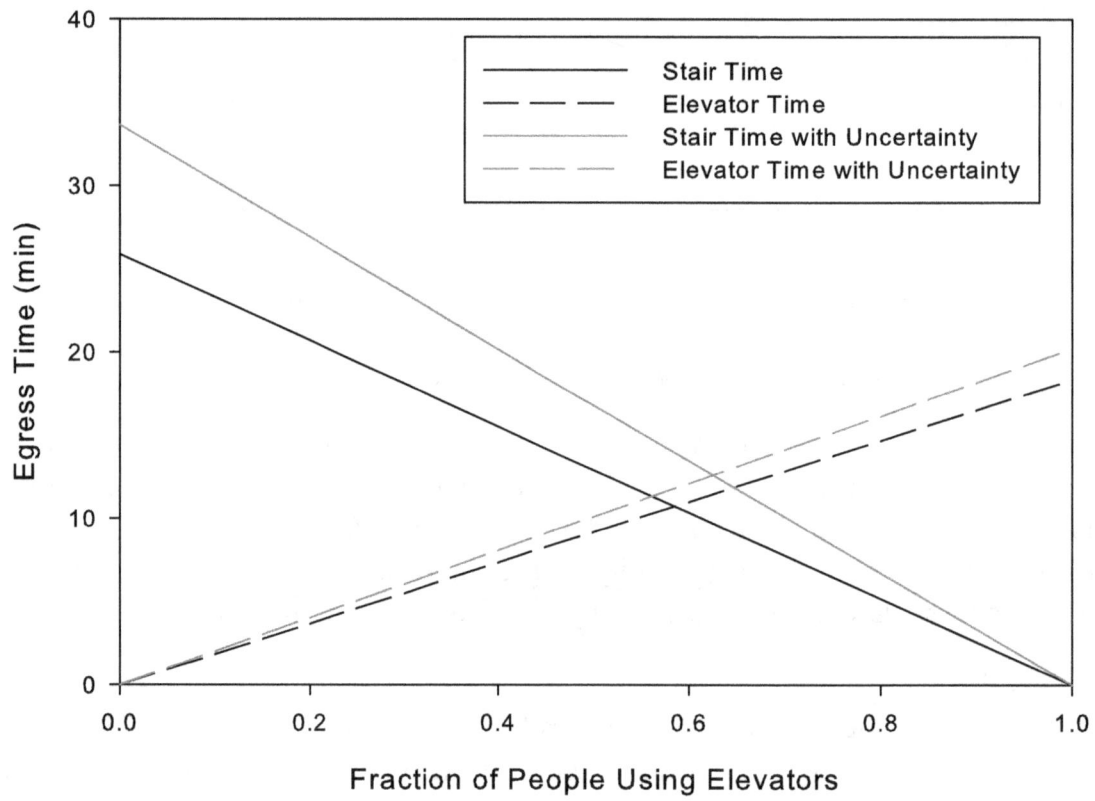

**Figure 28. Considering design margins in the combined analysis**

Note that this changes the ideal fraction significantly toward a higher fraction using the elevators. However, since the data is so limited on what the design margin should be, this should just be considered an example.

44

## 5.3 Phased Evacuation with Combined Stair and Elevator Evacuation

In a phased evacuation, as described in section 3.2, only the two floors above, the fire floor and the two floors below are evacuated. From the phased evacuation data, assume the building has 2 stair shafts with 2 flights of stairs per floor each 1219 mm (48 in) in width and 10 elevator cars that can each carry 10 people. Further assume that 50 % of the people use the stairs and 50 % use the elevators. The results are that it takes 11.8 min for 256 people using the stairs to evacuate and 14.3 for the same number to evacuate using the elevators. From Eq (25) and (26), the ideal fraction is 0.45 and the estimated egress time is 12.9 min. When using 0.45 as the fraction in the egress estimator the stair egress takes 12.9 min and the elevator takes 12.9 min. Adding this data to the previous data gives a new ideal fraction of 0.46, which is essentially the same as the original estimate. The results can be seen in Figure 29.

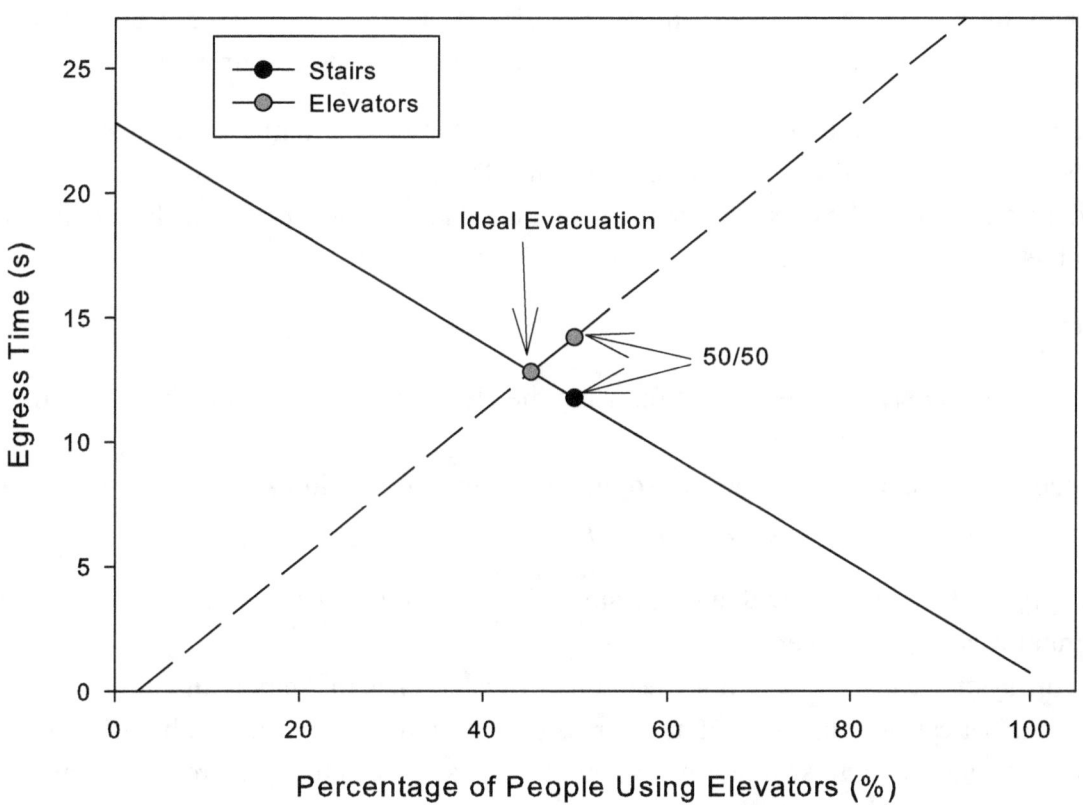

**Figure 29. An analysis of phased evacuation of 5 floors with both elevators and stairs**

# 6   Summary and Future Research Needs

This analysis of the combined use of stairs and elevators in an evacuation shows that using both can reduce overall egress time. NIST's model of elevator and stair egress, the Egress Estimator, was used to do the analysis.

## 6.1   Summary

First, the sub-model of the stairs was examined to better understand the impact of changing different parameters. Three main parameters were examined, including the number of floors, fraction of population using the stairs, and number of stairwells used.

 As expected, wider stairs or more staris with the same total stair width lead to decreased egress times. Because more stair width is available, the density decreases. When comparing the 1219 mm-wide stairs and the 1118 mm-wide stairs, the effects are nearly linear with the effective egress width and the total evacuation time. However the relationship between density and total egress time is not linear.  For the two 1905 mm wide stairs, the total egress times are slower than the three 1118 mm wide stairs. Despite the effective area being 30 % more stair width.

Elevators are more complex systems than stairs. Parameters that were examined included the number of floors, the fraction of the population using the elevators, the capacity of the cars and the speed and acceleration of the cars. Unlike the stairs, there are nonlinear relationships between all of the parameters and egress time.

Findings include

- The relationship between the population using the elevators for egress and the egress time is linear.
- The total egress time increases as a quadratic of the number of floors.
- Total egress time decreases as $1/J$, where $J$ is the number of elevator cars.
- Total egress time decreases with a shape similar to an inverse relationship to the maximum capacity of the cars.
- When the system capacity, which is the number of cars times the maximum number of passengers per car, is held constant, the egress time with respect to the number of cars continues to have an inverse relationship, but the egress time actually increases linearly with respect to the capacity of the cars.
- Changes in elevator velocity and acceleration have little real impact except for the upper floors of a building.
- Breaking a buildings elevator system into zones without increasing the speed of any cars improves the egress time by improving the efficiency of the system, in other words, cars spend less time idle.

A method is provided that makes use of the linear relationship of egress time to passenger load in both the stairs and elevators to determine an estimate of the ideal fraction of evacuees that should use the elevators instead of the stairs. Uncertainty of the estimated egress time is also addressed based on an uncertainty of what fraction of the population uses the elevators. Finally, the method is used to show that even in a phased evacuation of just the affected floors, the evacuation can be improved by having a fraction of the evacuees use the elevators.

## 6.2    Limitations of Study

As with all studies there are limitations to the applicability of this work. The first set of limitations is the limitations of the Egress Estimator model [11]. The model is primarily a screening tool and currently only has limited validation. Some phenomena not currently included in the model are:

- Mobility challenged evacuees
- Counter-flow
- Uneven distribution of the population among floors
- Different demographic distributions of the population
- Stairs with differing designs in the same building
- The effects of smoke and/or fire
- The effects of building damage.

The second set of limitations is the finite number of designs studied. While the parameters that were studied were investigated over a wide range of values, they were still a small subset of design parameters for an actual building design. Some other design parameters may include

- Different elevator control algorithms
- Asymmetrical distributions of populations
- Differing elevator system designs (for example the use of sky lobbies)
- Varying floor plans and stair placements

The final set of limitations is the limited human behavior included in the model. The issue of what fraction would use the elevator was mentioned in section 5.2, but it is certainly more complicated. Other issues concerning human behavior include

- Different demographic populations
- Varying pre-evacuation time
- Interaction with smoke, fire and/or building damage
- Training and education
- Clear alarms and directions

While care does need to be made generalizing the results of this study, the basic conclusion holds that a combination of stairs and elevators as part of a evacuation system can significantly reduce egress time.

## 6.3 Future Research

There has been and continues to be a lot of interest in combining stairs and elevators into one egress system. This study is meant to be a step toward that goal but there is still research to be done. That research splits into two broad areas.

- First and foremost is the need for data on building evacuations, especially with elevators. The NIST project on collecting fire drill data is a useful resource but other sources of data would be helpful. There is little if any data on elevators use in evacuations to compare against model predictions. Elevator data is clearly the most pressing need on the experimental side.
- The model needs continued improvement. More validation work needs to be done to better understand and reduce the differences between Egress Estimator predictions and actual data. Once data is available validation work with the elevator sub-model also needs to be performed. The Egress Estimator needs to be made more flexible by allowing the fraction of the population using stairs and elevators to vary from floor to floor and to allow for different elevator systems including systems that use sky lobbies.

The question of using both stairs and elevators in a building egress system is an important one. This study addresses some of the issues involve but there is a significant amount of research yet to be done.

# References

1. Hall, J. R., Jr, "The Total Cost of Fire in the United States," National Fire Protection Association. Quincy, MA, 2011.
2 ."Design and Construction of Building Exits", National Bureau of Standards, Miscellaneous Publication M151, October 1935.
3. London Transport Board, "Second Report of the Operational Research Team on the Capacity of Footways", London Transport Board Research Report No. 95, August 1958.
4. Fruin, J.J. Pedestrian Planning and Design, Revised Edition). ElevatorWorld, Inc., Mobile, AL, 1987.
5. G. Proulx. "Movement of People: The Evacuation Timing." Chapter 3-13, pages 3-341 – 3-366. In *The SFPE Handbook of Fire Protection Engineering*. Society of Fire Protection Engineers, Bethesda, MD, third edition, 2002.
6. Lord, J., B. Meacham, B. Moore, R. Fahy, G. Proulx "Guide for evaluating the predictive capabilities of computer egress models." GCR 06-886, National Institute of Standards and Technology, Gaithersburg, MD, 2005.
7. Peacock, R.D., B. L. Hoskins, E. D. Kuligowski, "Overall and Local Movement Speeds in Buildings Up to 31 Stories." In *Pedestrian and Evacuation Dynamics 2010*, ed. R. D. Peacock, E. D. Kuligowski, and J. D. Averill. New York: Springer, 2011.
8. Hoskins, B. L., "The Effects of Individual Characteristics and Group Dynamics on Egress on Stairs." PhD Thesis, University of Maryland, College Park, 2011.
9. Nelson, H. E., and Mowrer, F. W., "Emergency Movement." Chapter 3-14, pages 3-367 – 3-380. In *The SFPE Handbook of Fire Protection Engineering*. Society of Fire Protection Engineers, Bethesda, MD, third edition, 2002.
10. Klote, J.H., An Overview of Elevator Use for Emergency Evacuation. CIB-CTBUH Conference on Tall Buildings. Proceedings. Task Group on Tall Buildings: CIB TG50. CIB Publication No. 290. October 20-23, 2003, Kuala Lumpur, Malaysia, Shafii, F.; Bukowski, R.; Klemencic, R., Editors, 175-185 pp., 2003.
11. Reneke, P.A., Peacock, R.D., Hoskins, B.L., Simple Estimates of Combined Stairwell / Elevator Egress in Buildings. NIST Tech Note 1722.
12. Kuligowski, E.D. and Peacock, R.D., "Building Occupant Egress Data," Report of Test FR 4024, Natl. Inst. Stand. Technol., Gaithersburg, MD (2012).
13. Klote, J.H., "Method for the Calculation of Elevator Evacuation Time," J. of Fire Protection Engineering, Vol. 5, No. 3, 86-96 (1993).
14. NFPA 101, Life Safety Code, 2012 Edition. National Fire Protection Association, Quincy, MA (2012).
15. Pauls, J., "Movement of People." Chapter 3-13, pages 3-263 – 3-285. *The SFPE Handbook of Fire Protection Engineering*. Society of Fire Protection Engineers, Bethesda, MD, third edition, 2002.
16. "Elevatoring Commercial Buildings," Chapter 10 in *The Vertical Transportation Handbook, Third Edition*, John Wiley & Son, Inc. (1998)